BEYOND THE OIL ERA?

Arab mineral resources and future development

BEYOND THE OIL ERA?

Arab mineral resources and future development

M.R. KHAWLIE

MANSELL

LONDON AND NEW YORK

First published 1990 by **Mansell Publishing Limited,**
A Cassell Imprint
Villiers House, 41/47 Strand, London WC2N 5JE, England
125 East 23rd Street, Suite 300, New York 10010, USA

British Library Cataloguing in Publication Data
Khawlie, M.R.
 Beyond the oil era?: Arab mineral resources and future
 development.
 1. Arab countries. Mineral resources
 I. Title
 333.8′5′09174927

 ISBN 0-7201-2040-3

Library of Congress Cataloging-in-Publication Data applied for

Printed and bound in Great Britain by
Biddles Ltd, Guildford and King's Lynn

To my wife, Saüsan Yamut,
for her sacrifice, sharing and enduring our presence
in Beirut, a city that once could have shown
Arab development at its best.

CONTENTS

FIGURES

TABLES

ACKNOWLEDGEMENTS

This book is the result of interest that has been aroused in me by several centres and individuals in the Arab World. Some have prompted me positively, others because of their negative attitude towards proper development. In all cases I bear full responsibility for the contents and the approach.

The Arab Organization for Mineral Resources has undoubtedly improved Arab views on co-operation with respect to these resources. Another important body that enlightened my views on Arab development is the Economic Committee for Western Asia. Many geological surveys in different Arab countries have likewise contributed, although to different extents. The Arab Geological Association has played a role, and I hope they will become more involved.

Many study centres, prominent among which is the Arab Unity Studies Centre in Beirut, have consistently made people aware of the necessity for more co-operation among the Arab countries in the hope of achieving better development. That will come through the optimum use of Arab resources, both physical and human.

The Lebanese National Council for Scientific Research, the American University of Beirut Research Board and the Geology Department have always supported my research, even under the most difficult situations that Beirut passed through. I am grateful to Mr Ali Monzer of the National Council for Scientific Research, Dr Ziad Baydoun of the Geology Department and several colleagues who helped me in many ways. Many of my students have added to my interest, including Mr V. Khariamirian, Ms F. Attiyeh, Ms L. Khalaf and Mr F. Haddad.

My appreciations extend to many who have facilitated the logistics of my work including my sister Mrs Thana Taymani, Mr Patrick Smith, Mrs L. Deeb and Mrs H. Nisr-Abdel Sater, both of whom typed the manuscript, Mr Maroun Ijreiss, who draughted the illustrations, and Mr John Duncan, who did an excellent job in editing and seeing the book through to its final production.

INTRODUCTION

The Arab economy is on the whole based in the oil producing and exporting countries. In the aftermath of the considerable decrease in Arab financial revenues, and hence Arab influence, from the declining oil sector, Arab states have begun to reconsider their economic policies and the effect on Arab national and regional security. There has been a decrease in Arab financial support to other countries, recurrent economic recessions and stringent financial policies. The effect has been reflected in reduced Arab co-operation, emigration of trained labour and more Arab countries having to rely on foreign markets. The exposure of the Arab economy to a fluctuating world economy led to a marked decline in Arab economic growth after the world recession of the early 1980s.

The Arab Gulf Federation of Chambers of Commerce and Industry in September 1987 reported a drop in financial resources affecting growth in all sectors. The major concern of those in the Gulf was the lack of liquidity. The notable example given was that of the Arab financial giant, Saudi Arabia, which spent $20 billion (all prices are given in US dollars) on construction in 1986, but was expected to reduce this by 10 per cent annually to reach $14 billion by 1990. Many construction companies have had to close or break previously signed contracts.

This grave situation results from Arab dependence on one major income source, oil. Although an economic revival occurred in the developed and some of the developing world in 1983–84, Arab countries, and notably the oil-producing ones, did not share in it. On the contrary, the Arab world experienced a reduction in growth rate of 3 per cent of GDP (AMF, 1985).

The oil price crisis in 1986 was a turning point for the oil industry (OAPEC, 1987). The negative effect on the Arab world was extreme, but prices stabilized after OPEC reinstated regulations for co-ordination among member states. Nevertheless, the unified Arab economic report (AMF, 1986) warned that although OPEC could stabilize oil prices at around $15 per barrel, this price would be almost 50 per cent lower than that in early 1985.

The continued fall of oil revenues eventually had a positive effect on

1

Arab countries. They became more careful in their spending, started redefining thier priorities in the use of resources, and tried to follow more logical paths of co-operation that would lead to proper long-term development (AMF, 1986).

The question is, what resources other than oil does the Arab World have? The answer is minerals, not only because Arab mineral wealth is extensive but also because integration of mineral resources into the everyday productive cycle for development will lead to industrialization. During the past 20 years many people have recognized the potential mineral power of the Arab World. Unfortunately, Arab authorities have been blinded by the radiance of black gold.

Arab countries have seen many advances in growth but not in real development. Oil revenues are still the driving force in Arab progress, but the approach is short-sighted and the results are limited. The alternative force is diversity, and the optimum channel is Arab mineral wealth. The Arab Organization for Mineral Resources has now been established,[1] and the oil producing and exporting countries of the Arabian Gulf met in Kuwait in 1986 at the First Conference on Indigenous Raw Materials and Their Industrial Utilization in the Gulf Region. The Conference recommendations reiterated the need for a strategy on mineral resources, specified that priority should be given to mineral resources and recommended that information should be centralized. This book examines these and other issues related to Arab mineral resources, along with the role they might take in development of the Arab World.

Development is a multi-faceted and complex, continuous process. Whether in an advanced country or a developing one, the significance of the process is enormous, but the emphasis varies. In developed countries it takes such forms as improving technical efficiency, increasing services, upgrading standing practices, etc.; in developing countries it has to do with building up the basic structure, both the superstructure and the infrastructure. Mineral resources are the one item that, if properly utilized, can push a country's economy, industry, trade, agriculture and overall well-being to high levels. Advanced countries have reached this stage, whereas the others, including the Arab countries, have not.

The Arab World is a typical bloc of Third World countries trying to develop (see Figure 1). This book explores their natural mineral resources and the role these will play in the complex process of transforming the Arab countries into a developed complementary block of coherent regions. The exploitation of mineral resources goes hand-in-hand with other socio-economic and physical aspects, but this book emphasizes only what directly affects the mineral resource sector for general Arab development.

The first part surveys mineral resources in the Arab World, showing current trends and the efforts being put into this vital sector. The second

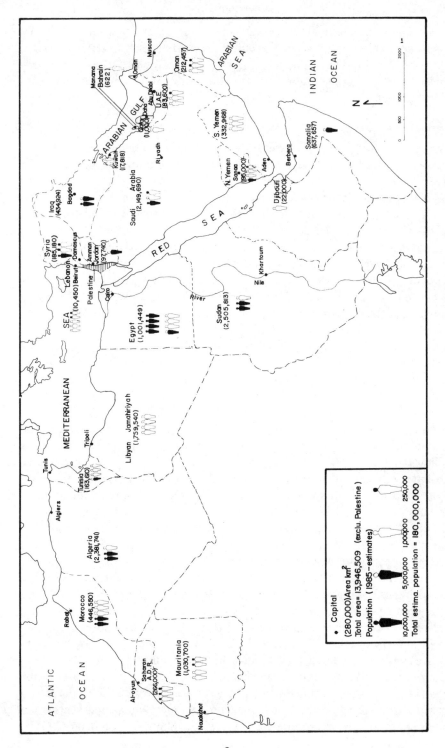

Figure 1. Countries of the Arab World, with areas and populations.

3

part deals with the position of mineral resources in the social structure, together with the related policies and financial factors influencing mineral resource development. The third part looks into the future, proposing three alternative paths along which the Arab World can proceed in its development.

Mineral Resources and Development

The economies of Arab countries are dependent on mining, including mineral resources in addition to oil. However, the mining sector in the Arab World is quite backward. As stated bluntly by Lenti (1986), the exports of less developed countries have been responsible for their patterns of backwardness, because they continue to export raw materials rather than manufactured goods.

The international mineral scene is geographically limited and fluctuating, reflecting such aspects as exploration, exploitation, exhaustion, technology and the need for intensive capital. Some Arab countries have benefited by selling some of their mineral resources as raw materials (e.g. phosphates from Morocco, mercury from Algeria, iron from Mauritania, sulphur from Iraq, potash from Jordan and oil from several Arab countries). This raw material based policy is short-sighted, because the benefit is neutralized by trans-national corporations' (TNCs) monopoly and unstable prices. A notable example is the early 1986 oil price crisis. Industrial Arab growth projections for the 1980s have already changed from an expected 9 per cent to 7.6 per cent (World Bank, 1979, 1984). The famous 1975 Lima Declaration, calling for 25 per cent of world industrial production to be located in developing countries by the year 2000, is now believed to be difficult to implement, although strategies for resource exploitation have been put forward (UNIDO, 1981*a*). This does not mean that no progress has been made, but the pace of industrialization is unsatisfactory. Policies greatly influence the use of existing resources and the development of new ones. Arab countries have to learn that resources are not just physical, but need a capable administration and interest groups to perpetuate them in the economy (Cody *et al.*, 1980).

Changes in the world are occurring at an unprecedented rate through technological innovation, better utilization of natural resources and developing population structures. As the rate of change is increasing, unless the Arabs catch up their resources will soon be neutralized. One Arab country imported salt from the USA when it could have exploited its own. Another imported iron from a distant country rather than from Mauritania, and could have made better use of its own iron. It is time for Arabs to strengthen their national institutions to make their operations as self-sufficient as possible and free from exploitation, as was stated in a conference on new directions in mineral development policies (AGID,

4

1977). There is increasing competition for resources for developmental needs, for finance and for skilled manpower. Policies must be flexible, well defined and implemented and, if need be, modified to optimize well-being in the coming decades (Khawlie, 1986*b*).

Arab countries are aware of the challenges relating to the development of their mineral resources, and of the great positive role that these resources might play in pushing forward their national and international interests. This was clearly indicated in the 1981 Amman Arab Summit Meeting and has been declared in several economic covenants (Dajani, 1982). Yet economic Arab integration is today at a critical stage, while the challenges are increasing. The oil era of the 1970s has brought new and different developmental trends, affecting the political, social and economic links among the Arab countries and resulting in overall negative structural features, notably reflected in polarization, separatism and differential growth (Zalzala, 1982).

Development Categories of Arab Countries

There is often confusion between growth and development. Growth normally should be linked to a specific sector or aspect, whereas development is more overwhelming and wider in context. Development requires basic changes in institutions, the economy, demography, technology and social and political aspirations, so that developing countries can build their own productive capacities to a level that assures continuously satisfactory standards of living (Sayegh, 1982).

There are several categorizations of developing countries. The report of the International Bank for Reconstruction and Development groups countries according to geography, thus merging all Arab countries into North Africa and the Middle East Region (IBRD, 1983). The annual World Bank Report classifies countries according to average income, thus separating Arab countries. The International Institute for Applied Systems Analysis (IIASA) has identified seven regions that describe the world, the regions differing in their economic development and availability of resources and, to a lesser extent, in geographical conditions. The Arab countries are placed at the sixth level of development (Häfele, 1980). Following a more localized view within the Arab World itself, the unified economic Arab report divides the countries into four groups (Sayegh, 1984):

1. Algeria and Iraq. Oil is the major contributor to GNP; considerable populations; diversified economic bases and availability of agriculture, water and minerals, geographical variability; economic–industrial output other than oil; available trained manpower.
2. Saudi Arabia, Kuwait, United Arab Emirates, Qatar and Libya. Oil

is the major commodity; minimal population; not much economic diversification; minimal trained manpower.

3. Bahrain*, Oman*, Syria*, Jordan, Lebanon, Egypt*, Tunisia*, Morocco and Palestine. These have moderate economic bases outside the oil sector (excluding Oman and Bahrain); a high degree of trained manpower. Those with asterisks are oil producers as well but not as major as groups 1 and 2.

4. North Yemen, Yemen People's Democratic Republic (PDR), Somalia, Djibouti, Sudan, Mauritania. These countries show the lowest growth rate and lag in socio-economic structure; this picture was just beginning to change in the late 1980s.

It is obvious from the above that classifications vary. Our concern is development as related to mineral resources. Dividing the Arab World using the UN geopolitical scheme (Khawlie, 1983*a*), the four regions are:

A. The Arab West: Algeria, Libya, Mauritania, Morocco, Tunisia.
B. The Nile Valley and the Western Red Sea: Djibouti, Egypt, Somalia, Sudan.
C. The Fertile Crescent: Jordan, Lebanon, Syria, Iraq (Palestine).
D. The Arab Peninsula: Saudi Arabia, the two Yemens, Oman, UAE, Qatar, Bahrain, Kuwait.

Table 1 shows their characteristic mineral resource features at present.

Recession or Recovery

Many world crises can be traced back to bad decisions or faulty policies. The so-called energy crisis of the mid-1970s was the result of accumulated faulty procedures by advanced countries which had shown their negative signs before but were not heeded. There is now a great deal of Arab literature calling on Arab governments to co-operate along the path to integration in order to absorb world fluctuations. But are they willing to listen?

Developing countries' growth in the 1980s was predicted to be considerably less than the average for the 1970s (IBRD, 1983). In addition, growth will probably be less than the increase in population. For non-oil commodities, industrial production in developed countries dropped by 4 per cent, thus reducing the demand for primary materials exported by many developing countries. Prices of minerals and metals were sharply lower than in 1981. There is an as yet undefined global economic slump affecting all commodities, and notably minerals (Landsberg and Tilton, 1986; Chender, 1986). The demand for copper, iron ore, aluminium and lead fell steeply, which concerns several Arab countries that are developing some of those resources: for instance, several Gulf states are establishing bauxite industries, and Oman, Jordan and Saudi Arabia are

Table 1
OVERALL MINERAL RESOURCE FEATURES OF ARAB REGIONS

	Region			
	A	*B* *	*C*	*D* †
Mineral availability				
Metallic (M), non-metallic (NM), selective metals (m)	M/NM	m/NM	NM/m	M/NM
Poor amounts (P), moderate (Mo), rich (R)	R/Mo	Mo/R	R/P	R/R
Geological investigations				
Surveys (GS), mapping (MP), other field investigations (OFI)	All	All	All	All
Industrial exploitation				
Raw materials (RM), processed raw materials (PRM), semi-processed (SP), first transformation (1st), complex product (Cp), minor scale (1)	RM PRM SP 1st Cp.1	SP 1st Cp.1	RM PRM SP 1st Cp.1	RM PRM SP
Trained manpower				
Satisfactory (S), not satisfactory (NS), moderately satisfactory (MS)	MS	S	S	NS
Current activity				
Very active (V), moderate (Mo), poor (P)	Mo-V	V-Mo	V-Mo	Mo-P
Public/government (Pb), private sector (Pv)	Pb	Pb	Pb,Pv	Pb,Pv
Investments				
High (H), medium (M), low (L)	H-L	M-L	H-L	H
Foreign (F), Arab (A)	A,F	F,A	A,F	A

* Applies especially to Egypt. † Applies especially to Saudi Arabia.

establishing copper industries. We do not want them to discontinue their efforts, but rather to co-operate among themselves, at least to secure a stable co-ordinated market.

Development decisions today are multi-objective in nature, involving economic, social and environmental aspects. Natural systems comprise renewable and non-renewable resources. The content, structure and function of these systems are assessed not merely for quality, but also to seek sustainable uses (Hufschmidt and Carpenter, 1982). Are the Arab countries exploiting their resources at maximum efficiency, i.e. producing finished complex goods from them? Are they trying to secure a sustainable use? The selling of their resources as raw materials does not suggest that they are.

In his address to the OECD parliamentary assembly of the Council of Europe in January 1985, the Secretary-General indicated that, compared to 1981 and 1982, output in trade and prices had improved but several worries still existed. What does the future hold for an international monetary system beset by large exchange-rate fluctuations? He stated that far-reaching changes are affecting the world economy, such as increasing interdependence and the growing role of the developing nations. He concluded that there are features characterizing a world of uncertainty, and therefore what is needed is a build-up of confidence based on co-operation (Paye, 1985). The process of proper industrialization of Arab mineral resources can benefit from the intentions, at least as expressed in seminars, of the petrochemical industries. The latter, in an OPEC/UNIDO/OPEC Fund seminar on co-operation held in Vienna in 1983, emphasized the mutual benefits that can be obtained by a rational integration and allocation of the various stages of the petrochemical industry according to regional resource endowments. Self-reliance is the critical notion and without Arab co-operation it will not be achieved and Arab development will be halted.

More recently, Al Wattari (1988)—the Acting Secretary General of OAPEC—stated that enhancement of the industrial sector is one of the most important aspects of Arab economic development because it is essential to the achievement of self-sufficiency. Although the manufacturing industries in Arab countries have expanded considerably, they depend heavily on the oil sector; hence, with the decline in world demand for oil, recent years have brought a downturn in many Arab economies. Adding to this, the competitive nature of the oil industry between Arab countries makes it essential that diversification and co-ordination be widely introduced.

Note

1. There are rumours (early 1990) that the AOMR may be merged with some other unit of the Arab League.

I

MINERAL RESOURCES OF THE ARAB WORLD

Our picture of the mineral resources of a region is undergoing a continuous change because of progress in the technology of extraction and assessment. The cost of each stage, from exploration through prospecting to marketing, will eventually decide the value of the mineral resource.

Using old methods of exploration and prospecting, a field party might discover an ore deposit but be unable to evaluate it, whereas with today's more sensitive techniques, this same deposit can be easily evaluated by type, grade, reserves and workability. Similarly, raw material available in large quantities will be worthless until it becomes a marketable product. If proper exploitation is to take place, adequate investment is needed.

Mineral resource assessment, particularly for development purposes, requires a complex structure of geological institutes with trained native personnel and researchers, proper mining experience, mineral dressing and manufacturing processes, adequate financial backing, managerial experience and, finally, a market for the raw or finished goods.

Few Arab countries have built the infrastructure necessary for such undertakings. Table 1 shows the current position. If the actual and potential mineral wealth of the Arab World and the history of geological mineral investigation in Arab countries are considered, it is clear that much more progress should have been made. In Table 2 (taken from Afyeh and Mansour, 1977; AOMR, 1981, 1985; Khawlie, 1983) the time–functionality data reflect the variability of geological mineral studies in the Arab World. Column a shows that 38 per cent have poor surveys, 29 per cent moderate surveys and only 33 per cent advanced geological services. Column b indicates that only one-fifth of the Arab countries have major income from their mineral resources. Finally, column c, which is probably the most important, shows one country in region A, one in region B, three in region C and none in region D having well trained native personnel. This drastic drawback has negative implications for the future development of Arab geological mineral involvement.

Several geographical spots in the Arab World were the centres of ancient civilizations that undertook some form of mining or metallurgy

Table 2
TIME-FUNCTIONALITY PERSPECTIVE OF ARAB MINERAL INVOLVEMENTS

Region	Country	*Start of geological activity*	*Start of mineral activity**	*Functionality geological institutes and exploitation*		
				a	*b*	*c*
A. Arab	Algeria	1800s	1840s	A	I	+ (3)
West	Libya	1930s	1940s	M	P	–
	Mauritania	1920s	1930s	P	M	–
	Morocco	1890s	1900s	A	M	+ (1)
	Tunisia	1880s	1890s	A	M	+ (2)
B. Nile	Djibouti	1940s	1950s	P	P	–
Valley,	Egypt	1800s	1820s	A	I	+ (1)
West	Somalia	1900s	1920s	P	P	–
Red Sea	Sudan	1830s	1840s	P	P	–
C. Fertile	Iraq	1900s	1940s	A	P	+ (1)
Crescent	Jordan	1900s	1920s	A	M	+ (1)
except	Lebanon	1920s	1930s	M	I	+ (3)
Palestine	Syria	1920s	1930s	A	I	+ (1)
D. Arab	Bahrain	1940s	1960s	M	P	+ (3)
Peninsula	Kuwait	1940s	1960s	M	P	+ (3)
	Oman			P	I	–
	Qatar	1940s	1960s	P	P	–
	S. Arabia	1920s	1940s	M	P	+ (3)
	UAE	1930s	1960s	P	P	–
	Yemen N.	1900s	1940s	P	P	–
	Yemen S.	1900s	1930s	M	P	+ (2)

* Denotes continuous activity (excluding times of war), in all cases begun by foreigners.
a. A, advanced; M, moderate; P, poor.
b. Contribution to economy: M, major; I, intermediate; P, poor.
c. + denotes trained native personnel: 1, good to 3, weak. – denotes need for improvement.

(Afyeh and Mansour, 1977). Notable are localities in Jordan (copper mining—3000-2000 BC); Tunisia (lead mining—Carthaginian and Roman periods); Saudi Arabia (copper, gold, silver—King Solomon and Abbasid periods); Sudan (copper, iron, gold—2000-1000 BC); Oman (copper—prehistoric, not well defined); Lebanon (iron—Roman period, and others not defined, notably during Phoenician times); Egypt (copper, gems, gold—Pharoahs; lead, sulphur—Romans); Morocco (iron, lead, copper, gold, silver, salts, zinc, antimony, asbestos—prehistoric and Middle Ages); Mauritania (iron—Middle Ages); Yemen (copper, iron, salts, gems—prehistoric and the Queen of Sheba's times).

A look at what Arabs have been exploiting during the past few decades does not show significant breakthroughs: most major mineralized areas were discovered earlier. The major mineral deposits being exploited by Arab countries in recent decades are ores of iron, copper, lead and zinc, phosphates, salts, and some construction and clay materials. Explorations and investigations have continued and new mineral discoveries have been made.

The 1970s

Most data on the Arab World in the pre-1980 period are erratic and inconsistent—or missing. The main reasons are the inefficiencies of the respective authorities, which are typical of developing nations. Production figures are most often available, and in a way are the most reliable, because the items produced are raw mineral materials, involving simple extraction and minimal processing. Other figures on imports and exports are more erratic because of their apparent complexity.

From the early to the mid-1970s, the total production of Arab mineral raw materials varied considerably (Tables 3 and 4). The materials that remained essentially constant were mercury and silver in the metallics, and insulators and fillers in the non-metallics. Those which increased between 10 and 20 per cent are cement, salts, cobalt, lead and iron. The

Table 3
PER CAPITA CONSUMPTION IN THE USA AND ARAB WORLD IN THE MID-1970s

	USA	Arab World
Metallics (kg)		
Iron/steel	547	26
Aluminium	22	~3
Lead	6.4	0.2
Zinc	5.5	0.3
Copper	10.5	0.25
Manganese	6.4	~0.2
Others	8.6	0.5
Non-metallics (kg)		
Cement	364	~121
Stone	4154.5	~593
Sand and gravel	3900	~975
Phosphate	141	10
Clays	223	~28
Salts	200	~25
Others	486	~40

Source: US Bureau of Mines (1979), ECWA (1977), Afyeh and Mansour (1977), Jabr (1980) and author estimates.

11

phosphates, zinc and copper production almost doubled, the manganese tripled and the sulphur increased six times. However, tin production stopped, while antimony decreased by almost 13 per cent and nickel by 30 per cent. This picture reflects the shaky character of the mineral sector in the Arab World, and can be explained by the poor and uncoordinated development policies of the Arab countries.

Figures on consumption are not readily available, especially those on *per capita* consumption that reflect how advanced a country is. The estimates in Table 3 were made on a comparative basis between US *per capita* consumption figures and those of developing countries (World Bank, 1979) and some given data on the Arab World. The notable feature is that on average a US citizen consumed from three to over 40 times as much as an Arab citizen did in the 1970s. The lowest gap is in construction materials (three to eight times), then mostly in the industrial non-metallics (up to 15 times) followed by a jump to the metallics (with copper at 42 times), with iron and steel bridging this (21 times). These figures are an indicator of the trends in Arab commodity consumption: capitalization in the construction sector, lagging to a great extent in the agricultural sector, and terribly deficient in the metallic industrial sector. It should be stressed that the bulk of the construction materials were and still are used by the oil-rich countries in building their infrastructures, and the iron is used mostly by the few Arab countries producing steel, prominent among which are Egypt, Tunisia and Algeria.

Exports of materials from the Arab World in the mid-1970s (Table 4) were dominated by raw materials, and in some cases all of the production was exported (AOMR, 1981). This applies to manganese, copper, silver, chrome, cobalt, mercury and antimony in the metallics; and fluorspar, barite, clays and many salts in the non-metallics. The other major raw material exports ranged between 60 and 75 per cent of production (including lead, iron, sulphur and phosphates). There was essentially no export of partly or fully processed materials, and the minimal amount produced was locally consumed. This applies to steel products, cement, phosphate fertilizers, refractories, zinc and some aluminium (the latter being processed using foreign raw materials in some Arab countries, notably Bahrain, Egypt and Lebanon, because of cheap fuel and/or technical labour).

The poorest data are on imports. This is not only because of the unavailability of data in the 1970s (ECWA, 1977; Afyeh and Mansour, 1977; Jabr, 1980; Arab Mining Company, 1981; Muharram, 1984) but also because the Arab countries' major imports are of manufactured or fully processed materials. There are other imports, but the emphasis in this book is on mineral matter that is produced in the Arab World and not adequately exploited or not channelled through the Arab market. On the other hand,

Table 4
ASPECTS OF THE ARAB MINERAL RESOURCE SECTOR IN THE 1970s

	Production	*Consumption*	*Exports*	*Imports*
Metallics				
Raw iron (10^6 t)	15.48	29.20	10.83	
Steel products (10^6 t)	1.51	2.62	1.31	1.92
Cobalt (10^3 t)	1.53		1.30	
Manganese (10^3 t)	119.25	19.05	165.00	0.45
Lead (10^3 t)	116.00	24.00	86.50	30.00
Zinc (10^3 t)	49.10	31.80	18.50	14.00
Copper (10^3 t)	13.60	22.40	10.60	3.00
Mercury (10^3 flasks)	10.95		10.40	
Silver (t)	33.67		32.00	
Nickel (t)	170.00			
Tin (t)	13.00			
Antimony (10^3 t)	1.90		1.84	
Chromium (10^3 t)	21.00		28.00	
Non-metallics				
Phosphate, crude (10^6 t)	17.95		15.75	0.03
Phosphate fertilizer (10^3 t)	801.00	239.50	510.00	24.00
Cement (10^6 t)	13.10	13.20	0.27	0.05
Clays (10^3 t)	88.00	60.00		350.00
Salts (10^3 t)	1.75	3.25	19.00	15.00
Sulphur (10^3 t)	2.10		2.10	600.00*
Insulators (10^3 t)	454.00			
Construction materials (10^6 t)	261.00	182.00		1020.00
Barite (10^3 t)	220.00		230.00	38.00
Fluorspar/feldspar (10^3 t)	43.00		62.00	11.00

Figures are for single years or for averages of years for which data are available for the years 1970–76.
Cobalt, manganese, lead, zinc and copper contributed very small amounts of processed products.
t = tonnes.
*Sulphuric acid.

the export–import relationship shows some contradictory features. This is because some items are available in one or more Arab countries and exported to foreign markets, while other Arab countries import them from foreign sources. It is obvious that in the 1970s there was no proper interactive Arab market.

An example of the programmes of the 1970s might be instructive. The Arab Potash Company was established in Jordan in the mid-1950s, with several Arab countries as shareholders, to extract various salts from the Dead Sea. Its initial capital was about $25 million. This project dragged along. By the mid-1960s the required investment became $100 million, and in 1977 it stood at $420 million; the programme was planned to be completed in 1982, which was when production started! (Muharram, 1984). The mineral resources related projects (Jabr, 1980; Khawlie, 1983*a*) involved raw and partly processed phosphates in Morocco, Tunisia, Jordan, Algeria, Egypt and Syria, with about 20 projects having a total annual production target of 50 million tonnes in the late 1970s and early 1980s. Iron ore mining projects were mostly in Mauritania, Tunisia, Algeria, Egypt and Morocco with potential projects in Libya and Saudi Arabia. Again the cumulative annual production target was around 50 million tonnes, which was expected to be achieved in the early to mid-1980s. Next in importance was copper mining, notably as raw ore material and partly processed, and limited to Mauritania, Morocco, Jordan and Oman, with a total annual production not exceeding 1.6 million tonnes of ore and only several thousand tonnes of processed copper by the early 1980s. At that time the metallic deposits of the Red Sea deeps were being investigated and a joint venture between Saudi Arabia and Sudan was established to extract copper, iron, lead and zinc. Other projects were for lead and zinc in the countries of the Arab West, adding up to 100 000 tonnes of ore; sulphur; chromium; manganese; cobalt; antimony; fluorite; potash; feldspars; barite; and asbestos. All except sulphur were for small amounts.

The Early 1980s

These times witnessed a definite change in the overall growth of the Arab mineral resource sector, but unfortunately in quantity but not equally in quality. There was diversification in raw mineral production and some commodities, and this was linked with one or another form of co-operation, although it was still minimal. Optimization of bilateral or multilateral resource exploitation was lacking. This is reflected by competition in the phosphate market, the duplication of aluminium production from imported materials and the export and import of products to and from foreign countries, instead of the establishment and definition of an Arab market; and by the heavy economic reliance on the export of raw materials instead of processed goods, which has the negative effect of near exhaustion of some ores (Table 5). Co-operation was more theoretical than practical in approach and implementation, e.g. establishing training centres, shareholding in certain projects, limited exchange of information through committees of a technical rather than enforcive nature, and the organization of meetings and symposia (AOMR, 1981; Khawlie, 1983*a*; ECWA, 1983; Muharram, 1984).

Table 5 shows the major resources picture in 1980–81. It is not claimed to be exhaustive because full data are not available: only material of Arab origin that is of significance is included, and the data for some countries, e.g. Iraq, Lebanon, Djibouti, Somalia and some Gulf states, are

Table 5
ASPECTS OF THE ARAB MINERAL RESOURCE SECTOR IN THE EARLY
1980s

	Production	Consumption	Exports	Imports
Metallics				
Aluminium (10^3 t)	380.81	960.00	132.95	235.76
Copper (10^3 t)	26.17	64.00	33.24	68.73
Iron/steel (10^6 t)	5.12	12.00	0.86	7.73
Lead (10^3 t)	245.61	128.00	163.66	50.05
Mercury (10^3 flasks)	25.00		10.76	49.78
Nickel (10^3 t)	0.13		4.29	3.12
Zinc (10^3 t)	168.23	96.00	59.78	60.65
Silver (t)	89.20		14.16	0.49
Gold (kg)	258.96			679.10
Chromium (10^3 t)	25.54		0.02	0.24
Manganese (10^3 t)	110.05	48.00	112.32	6.82
Tungsten (10^3 t)	3.20		0.02	0.17
Antimony (10^3 t)	1.18			
Cobalt (10^3 t)	6.27		0.02	0.05
Non-metallics				
Phosphate (10^6 t)	29.88	23.68	7.73	0.24
Nitrogen compounds (10^6 t)	2.24		1.23	1.06
Cement (10^6 t)	30.43	50.24	0.11	20.02
Clays (10^6 t)	1.41	12.80	0.07	0.98
Salts (10^6 t)	1.29	9.12	0.45	0.03
Construction materials (10^6 t)	3.81	237.00	0.30	1.92
Barite (10^3 t)	580.15		336.66	87.80
Fluorspar (10^3 t)	100.38		6.03	
Feldspar (10^3 t)	92.08		58.05	17.70†
Gypsum (10^6 t)	2.04		0.25	0.10
Sulphur (10^6 t)	7.13*		0.08*	1.54
Sodium/potassium compounds (10^3 t)	16.90		13.45	184.11
Abrasives (10^3 t)			0.46	71.07
Mica (10^3 t)	3.89		1.00	2.26
Asbestos (10^3 t)	0.33		0.01	94.44

* Exclude data for Iraq, which would make a significant difference.
† Feldspar *and* fluorspar.

incomplete (Khawlie, 1983a; ECWA, 1983; US Bureau of Mines, 1983). Aluminium now appears because of the importance of its developing industry, although this had the drawback of importing all its raw material. The other metals follow different trends. Copper, iron, mercury and zinc show an increase in production and export, but imports still far exceed exports. This again can be explained by a lack of proper co-ordination between the different Arab countries, as reflected in the almost total absence of a local market. Although lead and manganese have better export–import relationships, the former is more processed or semi-processed, while the latter is still raw and its production and export follow a decreasing trend. There has been a considerable increase in production of cobalt, but falling exports and rising imports. It should be noted here that, while the data may not be fully complete, they nevertheless aid understanding, sometimes despite their contradictory nature. The rest of the metals are not very significant or have stayed fairly constant.

The non-metallics show a much better picture, in terms of both quantity and quality (Table 5). There was an increase in production and export and, most notably, a diversification in materials and in processed or semi-processed products. Imports, however, were still very prominent. The export figures unfortunately include many imported materials that were partially processed and/or simply re-exported unchanged. This is a part of the services sector that started acquiring a larger share of gross national product, and is explained by port-tax regulations and cheap transport or cheap fuel in some Arab countries. The major materials showing a significant increase in production and export from the 1970s are fertilizers (phosphate), cement and salts, with imports slightly increasing or remaining minimal. Fluorspars and feldspars doubled production with a stable export–import relationship. The other major non-metallics (clays, construction materials, sulphur and barites) increased quite considerably in production, exports and imports. Table 5 shows several other materials contributing to the non-metallic industry of the Arab countries. Some are very major, such as nitrogen and sodium compounds, while others are minor, such as abrasives, mica and asbestos; gypsum had a better position than before, but should have been even more important.

A direct indicator of the growth taking place in the Arab countries is the increasing consumption figures for almost all materials and commodities, excluding construction materials (Table 6). The peak in consumption of construction materials in the 1970s was caused by the abnormal influx of money from the increase in energy prices, which led the oil countries to start building up their infrastructure at a great rate. This sudden expansion of construction was unprecedented and not properly planned, and so naturally in the 1980s it subsided. The other materials showed an average doubling or tripling in *per capita* consumption. Phosphate's very fast rate of increase reflects the expanding agricultural sector. The increase in consumption is seen as a positive trend, but the figures are still low compared to the USA, even using 1970s figures (Table 3). The point here is that the gap is getting even wider: there is 'growth' but not enough development.

Table 6
ARAB *PER CAPITA* CONSUMPTION IN THE EARLY 1980s

	Consumption (kg)	Change from 1970s
Metallics		
Iron/steel	75	3-fold increase
Aluminium	6	2-fold increase
Lead	0.8	4-fold increase
Zinc	0.6	2-fold increase
Copper	0.4	2-fold increase
Manganese	0.3	1.5-fold increase
Non-metallics		
Cement	314	3-fold increase
Construction materials	1485	Decrease
Phosphate	148	10-fold increase
Clays	80	3-fold increase
Salts	57	2-fold increase

It is essential, when evaluating the Arab mineral resource sector in terms of its developmental potential, to look at regional contributions and interactions. Table 7 gives figures for production, export and import for the metallics and the non-metallics on a regional basis using the geopolitical regions noted before (Khawlie, 1983a). Region A, the Arab West, can be considered as the sole producer of metals, excluding chromium and iron. The huge iron potential of Mauritania and Libya may tip the balance in favour of Region A, while the total production of chromium is insignificant anyway. The export figures are quite different from those of production. Some metals are not exported, implying they are used up locally and, as before, the figures include export and re-export data, some items being exported raw while others are processed to varying extents. The imports picture reflects a wider spread over the four regions. While Region A's metallic wealth is shared by the different countries, except Libya, that of Region B is mostly from Egypt (and partly Sudan, which if properly developed may provide a rich mineral resource in the Arab World), that of Region C is very minimal and Saudi Arabia is the backbone of Region D.

The non-metallics barite, fluorite and feldspar are produced in and exported from Region A alone. The remarkable position of the Moroccan phosphates is very well known, but the other regions are also rich in this item, and contribute slightly more than 20 per cent. Region B dominates the production of clays, salts, gypsum, sodium compounds, mica and asbestos. Region C's major contribution is sulphur (mostly from Iraq,

17

Table 7
PERCENTAGE OF TOTAL ARAB PRODUCTION, EXPORTS AND IMPORTS TAKEN BY EACH REGION (A–D), EARLY 1980s

	Production				Exports				Imports			
	A	B	C	D	A	B	C	D	A	B	C	D
Metallics												
Aluminium	—	35	—	65	0.84	34.4	0.6	64.15	22.89	2.71	5.38	69.01
Copper	100	—	—	—	81.36	0.52	2.11	16.0	53.06	7.24	9.41	30.28
Iron	20	52	9.3	10.3	36.75	6.4	1.31	55.53	43.92	3.76	11.36	40.94
Lead	100	0	0	0	98.54	0.44	0.22	0.79	46.98	21.66	12.66	18.69
Mercury	100	0	0	0	100	0	0	0	0.65	1.16	0.01	98.18
Nickel	100	0	0	0	98.78	—	0.14	1.07	43.89	1.31	0.45	54.34
Zinc	100	0	0	0	97.14	0.09	0.46	2.29	63.18	8.69	4.64	23.46
Silver	100	0	0	0	n.a.				n.a.			
Gold	96	4	0	0	n.a.				n.a.			
Chromium	0	100	0	0	—	—	100	—	41.90	45.64	5.39	11.20
Manganese	99.6	0.4	0	0	100	—	0	—	41.19	26.87	0.01	31.91
Tungsten	100	0	0	0	34	0	0	66	12.86	—	—	87.13
Antimony	100	0	0	0	n.s.	0	0	0	n.s.	—	—	—
Cobalt	100	0	0	0	—	—	—	100	5.55	—	—	94.44
Non-metallics												
Phosphate (raw)	79	2.37	18.57	—	21.38	0.41	78.19	0.02	(fert) 64.88	17.22	6.66	11.23
Nitrogen	6.71	23.17	14.76	55.34	n10.0	n1.0	n1.0	n88	61.57	35.69	1.69	1.04
Cement	43.92	11.90	36.61	7.56	n12.0	n1.0	n7.0	n80	11.83	14.68	2.11	71.36
Clays	7.61	89.77	2.60	0	52.01	3.20	37.89	6.88	raw 71.04 / proc. 21.15	17.99 / 5.53	3.63 / 3.76	7.32 / 69.53

18

Salts	33.84	46.92	10.93	8.30	84.04	11.02	4.0	0.76	19.65	0.29	5.66	74.39
Construction materials	9.23	13.78	n42	n35	19.68	0.78	75.27	4.23	9.85	1.07	8.23	80.84
Barite	99.63	0.36	0	0	100	0	0	0	44.73	0.30	2.77	52.18
Fluorite	99.46	0.53	0	0	100	0	0	0	50.57	39.50	7.40	2.51
Feldspar	96.22	3.78	0	0	100	0	0	0				
Gypsum	18.60	47.88	16.37	17.13	80.95	—	16.49	2.54	2.68	1.02	2.74	93.54
Sulphur									94.63	4.65	0.60	0.10
H_2SO_4									24.52	3.60	34.79	37.08
Sodium compounds	4.14	45.56	?	50.29	—	55.17	2.65	42.17	36.47	42.41	10.58	10.53
Abrasives	—	—	—	—	2.17	61.65	14.38	21.78	93.65	0.96	3.02	2.31
Mica	46.38	53.51	0	?	48.69	36.44	0	14.85	15.43	3.90	7.05	73.61
Asbestos	—	n100	—	—	7.14			n90	38.08	4.64	1.97	65.93

(The brace in the table groups the Fluorite and Feldspar export figures: { 50.57, 39.50, 7.40, 2.51 })

Because the very important data for Iraq are unavailable, production and export figures for sulphur compounds are not given.
Only region D produces processed clays.
Construction material figures are questionable because of non-standardization.
n.a., no accurate information; n.s., not significant; n, not very accurate information.

19

although no accurate data are available), construction materials and a major portion of the cement. Region D supplies nitrogen and sodium products, a considerable amount of construction materials and an uncertain amount of sulphur. The export–import figures do not reflect only materials of Arab origin, but also the re-export of imported materials to various extents. Region D is by far the major importer, exceeding the other regions in more than 50 per cent of the items listed.

The Arab import–export market of the early 1980s is shown for some important metallic and non-metallic minerals in Figure 2. All the metallic imports are from foreign sources, except for a few cases, notably iron and aluminium, which are transferred intra-regionally. For non-metallics, Region A depends to the greatest extent on foreign import, followed in a decreasing order by Regions B, D and C. The only item imported solely from other Arab countries is raw phosphate, while construction materials, clays and salts have both Arab and foreign sources. Exports from Region A are mainly to foreign countries, except iron and construction materials, which are exported within the region itself. The exports of the other regions are more Arab based and inter-regional.

The major function mineral resource projects of the early 1980s were not different from those of the 1970s (Khawlie, 1983a). They consisted of: phosphates (the same countries as before, plus Iraq), with a few more projects and production capacity of around 60–70 million tonnes; iron, which was much as before, with Libya entering production; copper, mercury, zinc and lead, all of which were expected to have become more important but were restricted to the same countries as in the 1970s, i.e. copper in Mauritania, Morocco, Jordan and Oman, and the others mainly in Algeria, Morocco and Tunisia. The projects that were intended to be productive by the early 1980s (the iron of Morocco, Tunisia and Libya, the copper of Mauritania, Jordan and Oman) did not work out as planned in time, capacity or even in quality (processing). Other potential projects, for extraction of copper in Saudi Arabia, Sudan and Yemen, and extraction of lead, zinc and other metals from the Red Sea deeps, were also incomplete.

Table 8 is taken from Khawlie (1983a) and compiled from the national papers presented by the Arab countries at the Fourth Arab Conference for Mineral Resources in 1981. When compared with previous data, it shows the very basic shortcomings in Arab mineral resource planning and exploitation.

Table 8 gives a good picture of projects that were to be implemented or were at an advanced stage of evaluation for exploitation. Of those listed, only the sulphur project in Iraq was completed and the others were delayed very badly. The Fifth Arab Conference for Mineral Resources, held at Khartoum in 1985, gave information of what happened next.

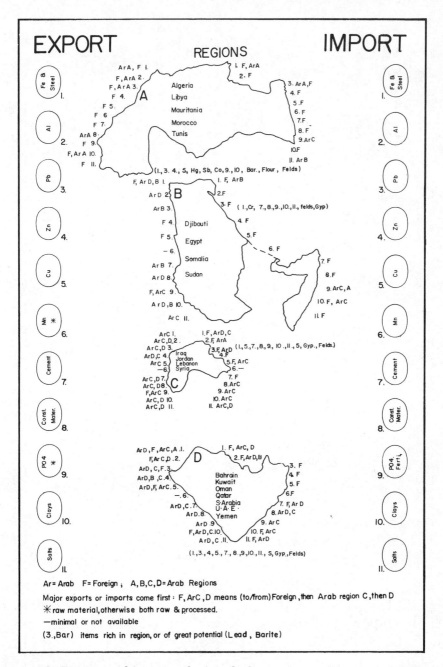

Figure 2. Exports and imports of minerals, by region, in the early 1980s. This figure shows the destination or source of each of the minerals for each region.

Table 8

IMPORTANT PROJECTS ANNOUNCED IN 1981 AT THE FOURTH ARAB CONFERENCE ON MINERAL RESOURCES

Country	Region	Manganese	Cobalt	Chromium	Sulphur	Fluorspar	Barite	Others
Algeria	A						135 000 (82)	U (84) 1200 kg, clays, Au
Libya								Salts, clays, gyps., U, const. mater.
Mauritania				Possible		Possible		Gyps, Au, Ag, U, Mo, Sa, W
Morocco		491 (80)	31 536		u.s.	221 000 (80)	340 000 (80)	Salts, Sb, Au, Ag, Hg, Mo, Sn, W, Ni
Tunisia						70 000 (90)	90 000 (90)	Salts, Hg
Egypt	B	Mid-1980		u.s.			50 000 (80)	Ti, salts, asb., Sb, Sn, U, Mo, Ni
Somalia								Gyps., Sn, Mo, REE
Sudan		u.s.		25 000 (80)		u.s.	Possible	Asb, kya, U, Au, Ag, W, clays, gyps., mica, Ni
Iraq	C				700 000			U, clays, gyps., salts, Asb., const. mater.
Jordan		u.s.			u.s.			Salts, feldsp., gyps., clays, U, const. mater.
Syria		u.s.		u.s.	u.s.			Clays, gyps., salts, Mg, const. mater.
UAE	D	u.s.		u.s.	u.s.			Salts, feldsp., gyps., clays
S. Arabia		u.s.			u.s.		u.s.	Extensive met./non-met.
Oman		(1980s)		u.s. (1980s)				Abs., Ni, const. mater.
Yemen N.			Possible					Assessment stage, very promising
Yemen S.					u.s.		Possible	Assessment stage, very promising

Figures are expected production in tonnes. Expected year of production in parentheses. u.s., under study for exploitation.

Arab Mineral Resources—The Current Scene

The February 1985 Fifth Arab Conference for Mineral Resources was attended by only 15 Arab countries, but also by several organizations and institutes. The main input was from the Arab Organization for Mineral Resources (AOMR), which presented a major working paper. This summed up the results of several investigations in the Arab World concerned with strategic minerals, precious materials, the economics of small-scale mining, and other mineral sector related aspects. The introductory remarks of the Secretary General of AOMR, which were the same remarks as were made in the previous four conferences, emphasized the fact that the Arab World is still very far behind in exploitation of its mineral resources. It contributes a tiny percentage of world production, and the majority of exports are still in the natural ore or raw state. Again and again the main working paper (AOMR, 1985) emphasized the increasing importance that mineral resources are acquiring in the developing world, including the Arab countries, in pushing the development process and in their direct link with industrialization. Equally, the paper makes the ever-repeatable call for co-operation, which is still minimal, among Arab countries in the mineral resource sector, from exploration through to marketing. This makes one wonder how many future conferences will repeat the same verses before basic functional co-operation takes place.

On a global scale, only the phosphates, sulphur, mercury, lead, zinc, iron and copper deposits of the Arab World are significant. The other resources, as shown in Table 9, are still not properly developed. The data of Table 9 reflect an inherent lack of proper information exchange and resource delineation in the Arab World. The maps on Arab mineral resources that finally appeared in 1987 (of restricted availability to the public) should serve to document the geographical and geological distribution, and in some instances the status, of Arab mineral resources. The AOMR, in co-operation with Robertson Research International, has produced these maps, which cover the Arab World at scales of 1:5 million and 1:2.5 million and show the economic importance of about 4000 deposits (Collinson *et al.*, 1987). It is a pity that, at the end of 1989, these maps are not yet public. In fact, their extremely high cost will make it difficult for many libraries at research institutes, not to mention individual researchers, to obtain them. This will not in any way serve anyone interested in Arab mineral resources for academic or industrial reasons, both of which are important as contributors to development of these resources. The underlying issue of information availability is again obvious.

The poor state of mineral development is explained by the unilateral approach followed by individual Arab countries in exploiting these resources. Most of the resources are not located in easily accessible places, and hence great financial and technical efforts are needed for proper development and benefit. It is worth noting that some Arab countries have fuel problems despite the Arab World having about 56 per cent of the world reserves of oil in 1987.

Table 9 gives a good idea of current Arab resources. Not many new finds have been declared and there is more concentration on exploiting what is already known about. Other 1980s data significant for developmental purposes, i.e. on production, exports, imports, and consumption, are not yet available in a systematic pattern that can be used. These data will probably not be available for some time, and then only through certain world agencies, such as the United Nations or World Bank. Only some production figures can currently be obtained (AOMR, 1985; UN Bulletin, 1985, 1989; United Nations, 1984*a*, *b*, *c*, *d*; Europa International, 1985; Wilcock, 1987; BGS, 1988) and these are shown in Table 10. The data of Table 10 are, as usual, lacking which is especially obvious when they are compared to data obtained during the Fourth Arab Mineral Conference held in Amman in 1981 (Khawlie, 1983*b*). However, they do give some useful indications. Algeria is becoming more prominent in the mineral sector, and the same goes for Egypt and Oman. Other Arab countries that should now acquire a better position than before are Saudi Arabia, Sudan, Yemen and Jordan. The first has several new finds and is revitalizing old mining prospects, especially for gold. The other three are faced with either financial constraints or political instability. A comparison of Tables 9 and 10 shows that many, but not all, the important minerals (the first three categories of Table 9) are becoming more widespread in different countries of the four Arab regions. Table 10 also shows some increased production of mineral commodities categorized among the less important groups of Table 9, i.e. aluminium, cadmium, chromium, potash, diatomite, mica, bentonite and talc. This might indicate that more serious efforts are being made by Arab authorities, but it is certain that we are still a long way from 'proper exploitation' of these resources in a framework of co-operation and uniformity.[1]

The production figures for two important industrial commodities, steel and cement, are indicative of proper exploitation. Table 10 shows steel production at three million tonnes, which is only 50 per cent more than in the mid-1970s. Cement production has almost doubled to a minimum of 25 million tonnes. If these are considered together with the index of industrial production (UN Bulletin, 1989) for mining, manufacturing and construction in major industrialized Arab countries, the picture is again of the Arab World lagging behind even many of the developing countries.

24

Table 9
IDENTIFIED ARAB MINERAL RESOURCES

	Amount (tonnes)	Regions (in order of importance)
High-category deposits (important on global scale)		
Mercury	25 000	A
Lead ore	70 million	A,B,D,C
Antimony	20 000	A,B
Cobalt	15 000	A,D,B
Iron ore	11 000 million	A,B,D,C
Phosphates	64 500 million	A,C,B,D
Barite	12 million	A,D,B
Fluorite	14 million	A,B,D
Sulphur	2000 million	C,D,B,A
Economic deposits (important on pan-Arab scale)		
Silver*	11 000	A,B,D,C
Zinc ore	66 million	A,D,B,C
Copper ore	600 million	A,D,B,C
Manganese ore	18 million	B,D,C,A
Construction materials	Extensive	C,D,B,A
Salts	3500 million	C,D,B,A
Silica	Extensive	D,C,B,A
Feldspar	Extensive	B,A,D,C
Gypsum	Extensive	B,D,C,A
Near-economic deposits (important on country scale)		
Gold*	350	A,D,B
Titanium	150 000	D,B,C,A
Tungsten	110 000	D,B
Tin	55 000	A,B,D
Clays	Extensive	B,A,C,D
Dolomite	Extensive	D,C,B,A
Asbestos	25 million	B,D,C
Magnesite	150 million	B,D
Subeconomic deposits (future moderate activity)		
Aluminium ore	350 million	D,B,C,A
Molybdenum	3000	B,D,A
Cadmium		A
Chromium	65 000	D,B,C,A
Potash		C,D,B
Wollastonite		B,D
Diatomite	Extensive	A,B,D
Mica	150	B,D
Marginal deposits (future small-scale activity)		
Platinum		B,D,A
Nickel		D,B,A
Bentonite		A,D,B,C
Talc	50 million	B,D
Kyanite	550 000	B,D
Perlite		B,A,D
Minor deposits (far future activity)		
Beryllium	1000	B,D
Zirconium	13 000	B,D
Vanadium	44 000	B,D

* Saudi Arabia has large potential deposits of silver and gold; its projected gold exports for the year 1991 are 1500 kg.

Table 10

AVERAGE ANNUAL PRODUCTION FIGURES, 1985–88

Metallics	Region A					Region B					Region C						Region D				
	Algeria	Libya	Mauritania	Morocco	Tunisia	Djibouti	Egypt	Somalia	Sudan	Iraq	Jordan	Lebanon	Syria	Bahrain	Kuwait	Oman	Qatar	UAE	S. Arabia	Yemen N.	Yemen S.
Gold (kg)	a		a	a	0.87		a	a	268							1.7			a	a	a
Silver (t)	3.7		a	118.3			a	a	a										a	a	a
Copper: metal (10³ t)	0.2		a	17.53			a	a	a	a	a		a			17.07		a	a	a	a
Copper: refined (10³ t)							3.90									14.80					
Lead: metal (10³ t)	3.6			85.94	2.17		a	a	a	a									a		
Lead: refined (10³ t)	4.0			61.23	1.99																
Zinc: metal (10³ t)	1.3			14.37	5.33		a	a	a	a											
Zinc: slab (10³ t)	28.13			a																	
Aluminium (10³ t)	a	a		259	a		177	a	a	a	a	63	85	176		a		154	a	a	a
Cobalt (t)		a				a			a							a			a		
Iron: ore (10⁶ t)	3.37		9.16	0.19	0.31		2.05					a	0.19					a	a	a	a
Iron: pig (10⁶ t)	0.95				0.15		0.11					0.05	0.08								
Iron: steel (10⁶ t)	0.80				0.18		0.14				0.14	0.03	0.07				0.52		1.10		
Mercury (t)	775			595			a														
Antimony (t)	a																				
Cadmium (t)	112							a	a												
Strontium (10³ t)	5.4																				
Manganese (10³ t)	12		a	60	a		5	a	20	a	5		a			12			a		a

		C1	C2	C3	C4	C5	C6	C7	C8	C9	C10	C11	C12	C13	C14	C15	C16	C17	C18	C19
Metallics	Chromium (10³ t)	a	a	a	a	a	a	a	a	a	a	a	a	a	a	4	a	a	a	a
	Nickel	a	a	a	a	a	a	a	a	a	a	a	a	a	a	a	a	a	a	a
	Platinum																a			
	Vanadium					a	a										a	a		
	Titanium	a	a	a	a	a	a	a	a	a	a	a	a	a	a	a	a	a	a	a
	Tungsten	a	a	a	a	a	a	a	a	a	a	a	a	a	a	a	a	a	a	a
	Beryllium					a	a										a			
	Rare earth elements	a	a	a	a	a	a	a	a	a	a	a	a	a	a	a	a	a	a	a
	Tin	a	a	a	a	a	a	a	a	a	a	a	a	a	15	a	a	a	a	a
	Molybdenum	a	a	a	a	a	a	a	a	a	a	a	a	a	a	a	a	a	a	a
Non-metallics	Phosphates (10⁶ t)	1.13	a	20.95	5.56	0.76	a	a	1.0	6.3	a	1.6	a	a	a	a	a	a	a	a
	Sodium chloride (10⁶ t)	180	12	5.4	112	425	930	30	135	70	22	5	97	21	a	a	a	a	185	70
	Potash (10³ t)			460	8		260	a	a	a	647	a	a	a	a	a	a	a	a	a
	Asbestos (t)	60		38	a	3	a	a	a	a	a	a	a	a	a	a	a	a	a	a
	Barite (10³ t)	30	a	a	a	a	4.5	a	a	a	a	a	a	a	a	a	a	a	a	a
	Fuller's earth (10³ t)		a	a		a	18	a	a	a	a	a	a	a	a	a	a	a	a	a
	Feldspar (10³ t)	a	a	1		a	a	a	a	a	a	a	a	a	a	a	a	a	a	a
	Fluorspar (10³ t)			89		a	a	a	a	a	a	a	a	a	a	a	a	a	a	a
	Diatomite (10³ t)	2.7	a	a		a	a	a	a	a	a	a	a	a	a	a	a	a	a	a
	Gypsum (10³ t)	260	12.5	a	a	60	945	a	25	300	95	3	105	a	a	a	a	a	145	a
	Kaolin (10³ t)	11	a	a	a	a	115	a	a	a	13.5	a	a	a	a	a	a	a	90	70
	Magnesite (10³ t)			4.1		a	5	a	a	a	a	a	a	a	a	a	a	a	a	a
	Mica (t)					a	a	a	50	a	a	a	a	a	a	a	a	a	a	a
	Sulphur and pyrites (10³ t)	20*	14*	a	a	10*	a	a	180†	a	a	35*	44*	265*	31	37*	100*	1200*	a	a
	Talc (10³ t)		a	a	a	6.15	a	a	a	a	a	a	a	a	a	a	a	a	a	a
	Vermiculite (t)	154		a	a	a	a	a	a	a	a	a	a	a	a	a	a	a	a	a

a = presence of available ore, but no usable information. *Refined from hydrocarbons. †From refining and Frasch process.

Import–export financial figures for some Arab countries, one from each region, are shown in Table 11 (taken from United Nations, 1984*a, b, c, d*). It is clear that there is no proper interaction. The strong dependence on non-Arab imports and exports is very obvious, and this is common throughout the Arab regions. The figures for internal (Arab) exports for Jordan and the Jordanian–Moroccan internal imports are exceptions and do not follow the trend.

At this point it might be said that the import and export needs of the different Arab countries do not allow them to form a convenient market among themselves. This is really a questionable statement, because there is duplication, competition and waste of money and local raw materials in the commodities imported and exported. It is true that the Arab countries do not co-operate in their policies of development and resource exploitation. It is easy to imagine one or a group of Arab countries in one region or another capitalizing on a certain resource development or industrializing a certain commodity, while another country or group develops something else. The efforts should be complementary, with countries coming together to complete their supplies and demands. Again, some may say that this is a utopian dream. That may be so, but people have often talked about the size limitations of the intra-Arab market without looking at its real functional developmental needs. Does a developing Arab country, at the sixth grade on a global scale, need to import aluminium ore from a non-Arab source more than it needs to import salts from its next-door neighbour? Is it really true that the Arab market is not feasible, or is it that Arab development priorities are not well defined?

Table 11
EXPORTS AND IMPORTS FOR SELECTED ARAB COUNTRIES

	Jordan	*Saudi Arabia*	*Egypt*	*Morocco*
Exports				
World	528 341	79 124 880	3 120 195	2 058 599
Arab North				
Africa	12 401	933 103	72 550	69 167
	2.35%	1.18%	2.33%	3.36%
Arab Middle East	342 930	4 156 814	144 520	114 340
	63.70%	5.25%	4.63%	5.55%
Imports				
World	3 241 185	40 655 325	9 077 949	4 315 286
Arab North				
Africa	26 131	341 078	45 908	63 836
	0.81%	0.84%	0.51%	1.48%
Arab Middle East	901 355	1 472 804	222 933	908 208
	27.81%	3.62%	2.46%	21.05%

Figures are in $1000.

Table 12
SOME FUTURE MAJOR ARAB MINERAL RESOURCE PROJECTS

	Non-metallics							Metallics								
	Clays	Salts	Feldspar	Asbestos	Gypsum	Sulphur*	PO_4	Ti	Cr	Mn	Zn	Fe	Pb	Ag	Au	Cu
Mauritania	o o o	+ +			+ +	+	+ + +			+		+ + +	+	+	+	+ + +
Morocco	+ + +	+ +	+ +		+ +	+	+ + +			+ +	+ +	+ +	+ + +	+ +	+ +	+ +
Algeria	+ + +	+ +	+		+ +	+	+ + +				+ +	+ +	+ +	+	+	+
Tunisia	+	+ +			+	+	+ + +				+ +	+ +	+ +	+		+
Egypt	+	+ +	+ + +	+	+ + +	+ +	+ + +	+ + +	+	+ +	+ +	+ + +	+	+ +	+ +	+ +
Sudan	+ +	+ +	+ + +	+ +	+ + +	+ +	+ +	+ +	+ + +	+ + +	+ +	+ + +	+ +	+ + +	+ + +	+ + +
Jordan	+ +	+ + +	+ + +		+		+ + +	+ +		+ +		o o	o			+ +
Lebanon	+ + +						o o					o o o				
Syria	+ + +	+ + +	+	+ +	+ + +	+ +	+ +	+ +	+	+ +		+ +	o o			+
Iraq	+ + +	+ + +	+ +	+	+ + +	+ + +	+ + +		+	+	+	+	o o	+		+
S. Arabia	o o o	+ + +	+ + +	+ +	+ + +	+ +	+ +	+ +	+ +	+ +	+ +	+ + +	+ +	+ + +	+ + +	+ + +
UAE	o o	+	+	+					o o	o o		+				+
Oman	o o	+ +	+					+	+	+		+	+			+ +
Yemen N.	o o o	+ +	+	+	+ +	+ +	o o	+ +				+		o o	o o	+ +
Yemen S.	o o o	+ +	+				+	+ +				+		o	o	+

+ Satisfactory, + + moderate, + + + large resources. o Satisfactory, o o moderate, o o o large indications.
* Sulphur produced by oil refining is not included. Data are unavailable for Bahrain, Djibouti, Kuwait, Libya, Qatar and Somalia.

Future mineral resource projects

It must be admitted that several Arab countries have realized the importance of having technically sound and trained national personnel. I do not wish to deny that efforts are being made to advance the mineral resource sector. This study emphasizes, rather, that these efforts are unilateral, i.e. country by country, and so will always be lacking. The training of personnel and the opening up of specialized institutes for such a purpose are among the most significant future and ongoing projects, although they should be more consolidated and practical than they now are.

Since all Arab countries are trying to develop their mineral resources, future projects will be found throughout the Arab World for many mineral deposits. Copper projects are expected in numerous places, as are those for iron, silver, manganese, phosphates, salts, refractory clays, and so on. Table 12 shows planned projects for some major mineral resources. It is obvious that many, if not all, Arab countries will start their own mining development operations for this or that mineral deposit. The result will be identical under-resourced projects (because of the capital, technical and market intensive requirements) or projects without any future economic feasibility.

Would it not be better if, instead of weak unilateral material and human efforts being duplicated in various Arab countries, especially neighbouring ones, co-operative and thus functional efforts concentrated on particular projects, e.g. for copper in country X and for gold in country Y? Only when secure and positive results were obtained would efforts be oriented towards other similar projects. The framework of co-operation, whether technical or intermediary, could take several forms between or within countries or regions. This co-operation is a necessity from the exploration stage, through prospecting, mining and processing, to marketing. The interaction would ameliorate burdens that are too great for a single country to bear, but easily carried by a co-operative group of countries. These calls for co-operation are not new in the Arab World—what is really needed is the political will.

Note

1. The estimated *total* Arab production of metals is: gold, 600 kg; others, in 1000 tonnes, are: cobalt, 1; copper, 25; antimony, 2; alumina, 1450; silver, 35; zinc, 30; manganese, 130; chromium, 25. Table 10 does not show some of the important non-metallics such as silica sands, perlite, etc. and other rock-derived products, e.g. calcite, dolomite, volcanic ash, etc., purely because data are lacking, not because they are not being produced.

II

MINERAL RESOURCES AND ARAB SOCIETY

Mineral resources activities contribute directly to the welfare of a society in all their development stages. It is logical to expect an input, positive or negative, from the social structure into these developmental stages. This can be looked at as an interactive process in which efficient mineral activities push a society forward and, similarly, a well structured advanced society can increase the efficiency of its mineral resource exploitation.

In this respect humans are the managing power and stand to gain or lose from the results of the exploitation of natural resources. Since these resources are limited, the size of the population using them or affected by their development is a critical aspect, as is the technical ability. Science and technology, the research being undertaken into resources and the specialized institutes serving this research all enter the picture.

This chapter will examine this aspect of mineral resources in the Arab World, and will cover population growth, manpower in the minerals sector, and the roles of science, technology and training institutes.

HUMAN FACTORS

Population Growth in the Arab World

The population growth of the developing countries in the twentieth century has been a very decisive element in the imbalanced growth of their various economic sectors. This is particularly true for the Arab countries. The demographic changes which have occurred in the Arab World could have very serious consequences (Ta'mullah, 1982). Population growth was gradual in the 1950s and 1960s, increased in the 1970s to 3.9 per cent, and has levelled off currently, although it is still high at 3.3 per cent, one of the largest rates of increase in the world (World Bank, 1984). This population is unevenly distributed—extremely dense cities and coasts but sparse inner, mountainous and desert areas. It has created a negative actual and potential food sufficiency framework, which is both a cause and an effect of the ever-increasing waves of people leaving the countryside for the

31

cities. This in turn leads to the abandoning of those inner parts that usually contain some valuable mineral potentials, and hence to their deterioration. It is expected that the Arab World will contain about 265 million inhabitants in the year 2000, as compared to 160 million in the early 1980s (Al-Sammak, 1984). There will definitely be great difficulties in meeting the necessities of this huge mass if policies and plans are not formulated properly and in good time.

Manpower and the Mineral Sector

The nature of human resources in the Arab countries reflects certain of the countries' characteristics, such as low indigenous populations in some states and generally high birth rates and low death rates. The most significant characteristic is that Arab society is largely made up of youths: those aged less than 15 years constitute 45 per cent of the total population (Ta'mullah, 1982). It must be emphasized that the potential of these young people will not be realized as long as a proper, strategic, human resource planning policy is unavailable, as is currently the case. In a study on mobilization of Arab human resources, Shaw (1983) stated that their quantitative and qualitative distribution is the severest drawback. He saw the major problems as an excessive rate of construction with its associated drain of people from the countryside, poor agricultural progress, overgrowth of population, the slow entry of women to the productive process, and the lack of education and professional training.

In a situation of high economic growth, the Arab Middle East and North Africa (AMENA) region was employing close to 56 million labourers in 1985–86, an increase of 32 per cent from 1975. In 1975, 45 per cent of manpower demand was in agriculture, fishing and forestry, by 1985 this had fallen to 35 per cent. The mining and quarrying sector increased but stayed small compared to other sectors. The most drastic increases were in the sectors of utilities (86 per cent), construction (72 per cent) and services (52 per cent) (Serageldin *et al.*, 1983). Professional and technical manpower requirements in 1985 accounted for only 2 per cent of the total. Although this sounds small, it reflects an annual average compound growth of 9.6 per cent. Employment of non-nationals in the mining and quarrying sector changed from 36.8 per cent in 1975 to about 55 per cent in 1985. This reflects how underdeveloped the Arab countries still are in this field, and underlines the negative consequences that are inescapable if this kind of reliance on non-Arab manpower continues.

The manpower needed in mineral resource activities changes in quantity and quality with the scale of the resource and the stage of the activity, i.e. exploration, prospecting, mining, production, industrialization and marketing. The first two stages require highly qualified personnel, while the others require a large number of variably experienced specialized, professional, technical and ordinary trained labour. There are, of course, many other aspects directly affecting mineral resource manpower needs, which will be discussed later.

The 1974 survey of Arab mineral resource manpower listed a total of about 90 000 in all categories, with foreign labour accounting for 15 per cent of ordinary labour and 35 per cent of highly qualified personnel (Afyeh and Mansour, 1977). This difference persisted through the early 1980s, but with more foreign personnel on different levels in selective countries, notably in the Arabian Peninsula (Serageldin *et al.*, 1983). Manpower increased by almost 130 per cent to a total of about 140 000 (Khawlie, 1983*a*).

The AOMR study presented in Khartoum in 1985 indicated both quantitative and qualitative changes. The total now employed is slightly over 165 000, of whom 5 per cent are highly qualified, 20 per cent are at the medium level and the rest are at the ordinary labour level; the number of foreign personnel has decreased considerably and is almost restricted to the highest category. Although the number of Arab highly qualified personnel has increased by almost 55 per cent, this is still far lower than needed. (The numbers and percentages given are modified from those of AOMR as it gives 1983 figures, the modifications being obtained from national papers.) This situation is a reflection of the poor planning of Arab human resource use, which is the responsibility of the Arab authorities.

Technology and Mineral Activity

Many Arab countries are trying to update and re-organize their mineral institutes. They want to upgrade their services, and achieve better results to improve their economies and meet the welfare requirements of their societies. Technological progress and self-reliance are a must for achieving these objectives. The Arab countries are among those struggling to achieve progress by obtaining new technology, and I will now give an idea of its position in the Arab World.

Edwards (1983) gives a typical example of science and technology transfer and practices in the Arab World. He has experience of two technology projects in Egypt, one successful and the other a failure. The success occurred because: the project centre was established as an independent unit outside any bureaucracy; the personnel involved were of very high quality; the project had the full support of highly placed officials; the Egyptian director was completely in charge; and there was a great market need for the technology. Drawbacks were the absence of maintenance and repair teams and facilities, an uncertain quality 'benchmark' and poor communications among top management and other workers on the project. Edwards concludes that any technology transfer

33

is doomed to fail unless it establishes institutes and skilled national man-power to gear it and ensure continuity.

The Seminar on Technology Policies in the Arab States, organized by the Economic Commission for Western Asia (ECWA) and UNESCO in Paris in 1981, contained several papers on technology related aspects in different parts of the Arab World. Some of these were of a propaganda type, such as that by Bartocha (1981) on the role of the US National Science Foundation. Others dealt with varying topics and countries: the Algerian economy (Ben Shanho, 1981); technology and decision-making in finance (Corm, 1981); project evaluation (Bhatt, 1981); the trans-national companies (TNCs) (Michalet, 1981); industrial science in the Egyptian iron and steel sector (Maksoud, 1981); and a general paper by Oldham (1981), which drew on several global examples and ended with implications for the Arab states. Some of the implications had universal applicability, such as: the necessity of formulating a technology policy in association with economic and industrial planning; the need for policies to deal with the supply of and help create the demand for technical change; and the point that effective policies are those that refer to specific industrial sectors. Oldham also gave what he called 'the ingredients for developing Arab research capabilities'. These are: a group of policy makers aware of their society's needs for appropriate technology, that is, country and social structure specific; an institutional base to cope with research that is inter-disciplinary in nature, i.e. universities, science councils and societies that have continuous interactive links with industry; a cadre of trained researchers; a programme of research; and, of course, financial resources.

Since technology transfer is directly linked to industrialization, the Arab Organization for Industrial Development (AOID) published a study in its magazine on issues of technology transfer in the Third World (AOID, 1979). It blamed most of the problems of underdevelopment on Third World countries' being 'followers' and highly dependent on the advanced countries, and stressed that the countries were in bad condition because of the colonial legacy. The AOID study indicated clearly that technology was bought and sold as a commodity. As for any other transaction, there are characteristics of supply, demand, contracts, costs and risks. A more recent study published by the same organization claimed that the lesson has not been learned (Jalal, 1985). This study concludes that Arab efforts to build a sound and dynamic technological base have failed, because they have not created the framework for technology, which consists of science, research, experience, practice and institutional training. The magazine concentrated on Jordan, one of the Arab countries with a predominantly mineral resource oriented economy.

Technology transfer, and more importantly development of indige-nous Arab technology, has a further role as a means of increasing Arab co-operation. There are far-reaching social, economic, cultural and political consequences if Arab countries acquire indigenous technology through co-ordinated and well organized efforts (Zahlan, 1981). It is only through the co-operative achievement of high technology that the Arab mineral resources can be properly exploited. If co-operative technology is developed, the Arab market will stand a good chance of flourishing, with an internal supply and demand of mineral commodities that have been selectively and differentially developed in specific countries and regions for the use of all. This is not to claim that there will be self-sufficiency, but a better direction of efforts for the future.

In a more recent scientific gathering on this topic (Co-ordinating Conference on Technology in Muslim Countries—Development and Co-operation, Istanbul, 21–25 October 1985) the general tone of the conferees—many of whom were from Arab countries—reflected underlying discontent with the present technological position. Once again there were forceful calls for co-operation and supportive action to keep up with the growth of technological knowledge. Kettani (1985) indicated that the future development of the Muslim world does not look very promising on present trends. The lack of a sufficiently large, dynamic and imaginative scientific workforce and the brain drain are crucial in this respect. Many Muslim countries' economies are based on the export of their resources as raw materials. Although these countries have put great efforts into developing their research capabilities, the picture still remains very hazy, and the key to changing this situation is firm co-operation between the states (Kettani, 1985).

In 1985 an Arab conference on the application of science and technology recommended the establishment of an Arab centre for the analysis of science and technology transfer, but nothing came out of this except a feasibility study in co-operation with ESCWA (the UN Economic and Social Committee for Western Asia). The amount of money spent on technology transfer in the Arab World during the past 10 years or so is astronomical, at about $70 billion per year! Some studies done by ROSTAS (the UNESCO Regional Office for the Science and Technology of Arab States) indicated that the Arab World will need to establish three engineering colleges every year, together with 21 technical institutes, and to increase its average spending of $4 per year per person on research (compared to $8 by South Korea or $190 by Japan) (Aba Yazid, 1985; ROSTAS, 1985).

Two years later, in 1987, the Arab League Educational, Cultural and Social Organization (ALECSO) programme on Arab needs for total development was revealed in a conference in Damascus, where the secretary stated that two important features were unfortunately still prominent in the Arab World: there is no common agreed understanding, even within one country, on planning and development, especially in terms of the role of science and technology; all developmental projects relating to future strategies concerning the Arab World have been realized only as documents exchanged and speeches proclaimed at various meetings (Al-Olabi, 1987).

Mineral Resource Institutes and Training

It is by now obvious that the effective progress of Arab mineral development depends upon an upgrading of the manpower involved. It is necessary to create experienced and knowledgeable cadres at all ranks, i.e. highly qualified, medium level and ordinary labour. At present the high-calibre personnel are mostly of foreign origin, although the proportion of Arab nationals is increasing. However, the other two categories are more important. It is for these levels that specialized mineral sector related institutes and adequate and appropriate training programmes should be established. This is the immediate concern, and will be followed in the near future by acquisition of enough technology and know-how to train highly qualified experts, directors and scientists.

A problem in many Arab countries is their dependence on formal education at the expense of technical education. Most Arab educational systems are modelled on highly structured and academically oriented European systems, in which technical education has not been well integrated (Gilliam, 1987). Consequently these systems have failed to provide a labour force appropriately trained to fulfil national development goals. Technical education started to receive attention only after the early 1970s, and technical and vocational institutes still do not cover a wide spectrum of specialities, certainly not in the mineral industries.

All Arab countries now have some sort of geological service, consisting of official surveys, specialized departments in higher education or departments of ministries related to natural resources. There has been interaction with the geological bodies of advanced countries, such as the USA, USSR, UK, France, Germany and Italy. The interaction has been, and in several instances still is, one-way, i.e. transfer of technology, not implantation. Many Arab countries have established public or private sector institutes, companies and organizations specializing in certain mineral resource activities or including departments with this function. Some have tried to advance their technically oriented departments. Originally these efforts were the result of unilateral decisions and actions *within* individual countries (intra-Arab). It was only in the second half of the 1970s that inter-Arab institutes serving this purpose started to be formed, although recommendations for creating these co-operative bodies were made in the late 1960s. The first co-operative efforts were exclusively financial, or investment type, e.g. the Arab Mining Company, the Arab

Potash Company and many others. The build-up of inter-Arab technical co-operation was very slow in the 1970s, and involved foreign countries. The privately run establishments did not have well planned long-term policies, but short-term profit objectives, and the governmental geological surveys were administered and more or less held by foreign hands that had no desire to be replaced by nationals. Egypt, Iraq, Syria and to some extent Morocco, Tunisia and Jordan were among the few countries that did not follow this pattern.

The figures given earlier on human resources show definite progress in the 1980s, but at a rate much slower than is needed. Only two inter-Arab training institutes (in Morocco and Jordan) were formed for mineral resources, while the other institutes, in Egypt, Iraq, Morocco, Syria and Saudi Arabia, were limited in extent, scope and services, or were too general, and each served only one country. It is again clear that the institutes necessary for the creation of the required human resources, in both quantity and quality, do not yet exist.

The proceedings of the first Arab mineral resources conference in the late 1960s recommended the establishment of these institutes. The proceedings of the second conference in the early 1970s emphasized the importance of co-operation. The proceedings of the third conference in the mid-1970s talked about the two or three inter-Arab projects that were established. The proceedings of the fourth conference in the early 1980s commended the intended opening of the first inter-Arab training institute in Morocco. At about this time there was a breakthrough—the establishment of the Arab Organization for Mineral Resources (AOMR), an offspring of the Arab League. The AOMR grew in size and services, and is showing promising potential, but real problems are facing it, including the fact that Arab countries and authorities do not communicate properly with it, i.e. there is incorrect, inconsistent and missing information. Secondly, the AOMR does not have enforcive and binding powers on Arab mineral authorities—if it had, positive steps would have been taken in Arab mineral resource development. The second institute was established in Jordan in 1985. The fifth conference, held in February 1985, instead of being able to give concrete information, stated that data were lacking, inconsistent and non-standardized! This implies that although there are now two institutes, their functionality is still in question. Inter-Arab mineral resource information exchange is still very limited. The various Arab countries are still following unilateral policies, perhaps upgrading their national medium-level and ordinary labour manpower, but definitely not acting on the scale needed for a permanently growing inter-Arab mineral sector.

There is a basic requirement in the Arab World for a re-orientation, in many disciplines, of the technically specialized and higher learning institutes towards the training of medium-level and ordinary labour. This is particularly true in the mineral resource sector, as is obvious from several studies done in different Arab countries: on Jordan (Jalal, 1985), on Libya (Azzam, 1984), on Tunisia and in general (Bou Kamra, 1982), on Arab higher education and development (Ammar, 1982), on Arab human resources and the special role that institutes, universities and enterprises can play in pushing a unified development framework forward (Zahlan, 1981) and on mobilizing Arab human resources (Shaw, 1983; Serageldin *et al.*, 1983).

The re-orientation in the mineral sector has to take place on different levels, i.e. governments, regions and the totality of the Arab World. At a local level it is important to encourage vocational careers and strengthen needed institutes, to upgrade practising manpower, to train, and retrain after a certain number of years, miners, prospectors, administrators, middle-men, excavators, processors, marketers, transporters, and so on. This requires interaction between industry and institutes, not only the vocational ones, but also the universities. The motivations and expectations, the ranges of services and the roles of those taking part must all be clearly defined and implemented. It is imperative that if and when mineral development plans are made, they should be tailored to a workforce that can execute them. On a regional scale, the Arab regions and countries together should aim to create a balance between exploiting their mineral resources and serving the needs of their growing populations. This may require plans for demographic changes. Projects should be co-operative in planning and execution. This means creating an Arab common market, standardizing information exchange and strengthening transportation links.

As argued by Zurayk (1982), there are many demands for the Arab future, notable among which are: organized and advanced knowledge that requires proper institutes; the will-power to progress through self-evolution; and self-realization to creativity. The present is full of struggles and clashes that can be won only by full and positive mobilization of Arab resources—material and human.

POLICY FACTORS

Political decisions are the most significant factor affecting the process of development, especially if it involves several countries, and particularly so when one is dealing with a strategic component of the economy, such as mineral resources. Development supersedes the financial system to link with every aspect of the social structure; it is a social upgrading and a cultural revolution. This is why political will is vital. The Arab countries' only chance of achieving real development of their mineral resources is through the process of co-operation. In order for co-operation to be *real* and *effective* in the Arab World, the respective governments have to revolutionize their ideas about this process.

Contrary to many beliefs in official Arab circles, the co-operative process could and should be gradual but relatively rapid. Its hindrance would have negative effects not only on the total Arab economy, but also on the continuity of current growth. Arab co-operation has already been implemented in various sectors on a regional scale: the many cultural and economic ties in the Arab West; the co-ordination of various matters between Egypt and Sudan; the common market of the Arab Gulf states; and the intermittent (for political reasons) but vital relations between Iraq, Syria, Jordan and Lebanon, which come from prehistoric times. It should be clear that these ties are indispensable and do not need much policy formulation, which is why many are loose or shaky, and are affected by the smallest political jerk. The political decision thus required is one that will widen, upgrade and stabilize these ties and make them permanent, regardless of any changes of governments.

Mineral resource sector development is directly related to the industrialization of a country, which has to grow from within. Thus an 'endogenous industrialization strategy' is required (UNIDO, 1979). The Arab countries can become relatively independent from foreign control and manipulation of their resources in the future only through implementation of proper policies at state, regional and international levels. The exchange of foreign technology and other services for Arab resources should be done in such a way as to alleviate foreign hegemony over these resources. Just as other advanced countries have the right to secure their future supplies of resources, so too do the Arab countries.

There is no doubt that the 1970s witnessed an upheaval in Arab growth and (partial) development that left positive traces in social, economic and political areas. This would not have come about were it not for the critical change to Arab national control, especially over the oil resource (Sayegh, 1984). The prosperity that was enjoyed, however, will not come again for a long time, which is why policy planning for national resource exploitation is needed.

Policies: The Local Scale

Policies are not substitutes for projects, but the structure of policies can enhance or reduce project benefits. Mineral resource projects are capital, technology and human resource intensive; this is very important because the larger the anticipated investment programme, the greater are the potential losses from the wrong policies. The picture concerning technology has been discussed already. Policies on people, who are the most important resource in development, are almost non-existent in the

Arab World. The mineral activities of most Arab countries are pure reflections of the political regimes in their lack of links between the base and the top management. The Arab citizen has almost no say in the policies adopted. Since their independence, Arab countries have adopted pseudo-liberal, pseudo-social or strict monarchy-type governments that have all failed to build 'base–top' mutual relationships, most notably in policy-making. The reasons for this are many and varied, including social, economic, awareness and educational factors. The effects are similarly many and varied, and an imbalance in the social structure (Mu'awad, 1983). In the mineral resource sector this results in all mineral activities and operations being run by a 'ruling' management that dictates decrees to its labourers—there is no proper interaction, and therefore severe inefficiency.

For effective mineral development, the management of mineral enterprises should follow much the same course as the ruling authorities in Arab countries: there should be a real will to perform efficiently so that benefits prevail over all of society. The conditions needed to empower this performance must be created. Effective policy is linked to the totality of the regime or the enterprise, its structure, performance and mutual relations. Policy in the Arab countries has so far achieved different grades of failure (Assaf, 1982).

Several attempts have been made to relieve this problem. In Saudi Arabia the Deputy Ministry for Mineral Resources has taken large steps to implement development plans. The Deputy Ministry itself has grown and diversified, and is entrusted with responsibility for extensive programmes of mineral exploration and development, basic geology, geophysics, geochemistry and the provision of technical services supporting these activities. Mineral search is its prime function (DMMR, 1982, 1983) and much progress has been made, but the mining and processing is carried out by the private sector. This private sector is, unfortunately, mostly foreign controlled, which puts a big question mark over the Saudi policy.

Whereas the Saudi policy is limited, that of Egypt tries to go all the way through to the final market. This, of course, is mineral-selective, in that some minerals' development cycles are completely controlled by Egyptian hands (iron, phosphates, clays, glass, etc.), while others are not, but the important point is that the decision to gain full Egyptian control—in the short run—has been made. Many industries relying on mineral resources have formed following controlled growth of specialized technological sectors, institutes, investment, government operations, consulting and engineering firms, research and development. These have taken the regional and international picture into consideration (Maksoud, 1981). A few other Arab countries are following this path. Some are at an early stage, e.g. Algeria (Ben Shanho, 1981) and Jordan, others are at an intermediate stage, e.g. Morocco and Tunisia, and Iraq and Syria (GEGMR, 1984) are quite advanced in their national policies for mineral resource exploitation and development, relying almost entirely on national know-how. The Moroccan government is following a varied policy in which some enterprises are publicly controlled while others are privately controlled, and the financing is also varied. This is proving somewhat chaotic in some instances, but credit should be given for the establishment of 'corporate organizations' in the mineral resource domain. This has been suggested as a fruitful policy for other Arab countries, and notably would be beneficial in small-scale mining operations (Khawlie, 1986a). The distribution of Arab mineral resources in remote areas, together with the required size of operation, also makes the policy of small-scale mining and corporate groups a useful one for Arab countries.

Countries such as Sudan (DGMRS, 1984), the two Yemens, Mauritania and Somalia are very poor, and cannot but rely on external help. This is where regional and total Arab policy has to take part. It is far better that help should be channelled through Arab routes rather than, as usual, come with unacceptable conditions from foreign sources, notably the MNCs. That is why a *regional inter-Arab policy* is *essential*. In a case study of Sudan by the World Bank on resource allocation (Acharya, 1979), a strategy was proposed composed of elements as such developing the modern sector (meaning agriculture, and here extended to minerals) through foreign and domestic *private* sources! There is no way these private and foreign sources can help in the long run. The social benefits that a state should be planning to obtain from projects cannot be compatible with limited private concerns. The welfare of the Sudanese citizens cannot be compatible with foreign intervention by MNCs.

There are other policy forms in the Arab World, exemplified by the Gulf states, Tunisia to a limited extent and Lebanon. Here, especially in the last, the private domestic sector has almost full control of mineral resource activities, in some instances without any government intervention. In Lebanon there is an extreme liberal policy that may work in some parts of the economy, but that is bound to fail with regard to the development of mineral resources in the long term. In fact, Lebanon has a diversity of non-metallic minerals that have not yet been exploited as they should, and has an industry that is quite advanced but unstable. Its mineral resource development has been erratic in time (mining or related activities starting and stopping), in type (activities producing a certain product then shifting to something else), and in quantity and quality. Mineral activities are carried out on a day-to-day transaction process, and only a few have a defined structure. This is shown by the studies done by Khawlie (1983b) on the mineral industry of Lebanon, Khawlie and Hinai

(1980) on construction materials, Khawlie and Attiyeh (1986) on the ceramics industry, and Khawlie (1986*a*) on the feasibility of small-scale mining in Lebanon. All detail poorly developed mineral activity, if not a wasted mineral resource. Khawlie and Khanamirian (1985) investigated the iron–titanium black sands (placer) deposit in Lebanon, which is being illegally excavated as construction material! Why? Simply because the government policy has left resource exploitation entirely to the private sector. This is not all the problem, but there is no government policy related to development of the infrastructure and superstructure responsible for and facilitating mineral resource activities. This is true in both the public and the private sectors.

Policies: The Regional Scale

If local endogenous policies are necessary for growth, they are not sufficient unless they are coupled with proper regional interaction. There are many (actually all) Arab countries that cannot further their economic mineral development without some sort of integration with neighbouring countries. This applies to the totality of the mineral development operation, from early exploration to final marketing. A certain Arab country may have the technical personnel to do the job but may lack other aspects, such as finance, processing or the market. The lack of inter-regional co-operation explains the weakness of Arab mineral development. There are by now many inter-Arab projects, with the following characteristics (Murad, 1982): agricultural and foreign (non-Arab) oriented; a foreign dominated extractive industry; a consumer industry based on imports from foreign sources; a dominant capitalist 'follower' flavour; the lack of a complete productive framework and its infrastructural facilities; a rather low *per capita* GNP; a low production capacity; and insufficient and inefficient mobilization of resources for the social build-up.

The only effective mineral resource related inter-Arab enterprise established so far is the Arab Mining Company (more widespread than other offshoots such as the Arab Potash Company), which started in 1976 and within a short period proved itself to a considerable extent. In the first five years it was involved in over 12 mining projects in seven Arab countries, covering the mining, concentrating and processing of copper, iron, lead, silver, fluorspar, industrial minerals, radioactive materials and fertilizers. In that time it contributed about $850 million, and its total investments in these mining projects amounted to about $2 billion (Taher, 1981). Mr Th. Taher, the Director, has stated that the capacities of the Company are much larger, but that the co-operative response of different Arab countries has ranged between extremes of positive and negative. He indicated that Arab mining could show more positive results if there were proper co-operative policies. It should be emphasized that the rich Arab countries and those with a considerable background in mining have not really been co-operative. It seems that co-operation in Arab states is implemented shyly, or in the absence of any other choice. Perhaps this was to be expected, because authorities would be afraid of putting their eggs in the basket of a newly formed company, but even today, 16 years after the Company started, it is still finding it difficult to change the situation. The problem is again one of communication, and policy-making among the Arab countries, and is sad, because the Company has some very highly qualified Arab personnel with mining experience. Comparison with the successful Egyptian case, given previously (Edwards, 1983), suggests that the Arab authorities' bureaucracy and only partial support are working against co-operation. Moreover, it does not seem that they believe that their immediate interests will be served by, or that there is a market need for, Arab co-operation!

If a typical block of Arab countries that have been approaching each other rather closely—the Gulf states—is taken as an example, some points on Arab co-operation will emerge. Their co-operative development concentrates on the industrial sector and is overwhelmingly consumptive. It is a relationship of quantity not of quality, and time is working against it (Bsisso, 1983). In his study Bsisso suggested that these states must re-orient their co-operative development strategy as follows: invest their excess oil money largely within the Arab regional framework; create changes within the Gulf institutional bureaucracies; widen the scope of their interaction with other Arab regions; and eliminate the extravagant, superficial, consumptive character of their economies, changing to an internal, productive, strongly linked one.

The priorities for co-operative projects were set as recommendations of the Eleventh Arab Summit Meeting in Amman, Jordan, and emphasized industries related to basic needs. Many studies have been undertaken on opportunities for investment and joint ventures, and the sixth Arab industrial development conference gave the green light to 20 such projects. These included eight directly related to mineral resources: for iron, aluminium, sodium compounds, steel and the metals constituting it, phosphate and compounded fertilizers, glass, ceramics and other refractories. The information supplied included details of financial and manpower needs, but the most significant was on the geographical distribution of the different projects for each industry. There is a wide Arab coverage, the purpose of which is to further regional inter-Arab links through functional co-operation. Again, this means that the Arab World has the capacity, in personnel and other areas, to implement a push forward from selective growth to total development, but a decision is needed at the top levels.

Regional co-operation was examined by Sayegh (1983) in a study recommended by a group of Arab experts and requested by the Assistant Secretary-General for Economic Affairs of the Arab League. It included evaluation of 427 joint projects, of which 237 were purely Arab and the rest were Arab–foreign, covering finance, manufacturing industry, tourism and related services, agriculture, the extractive industry (the most directly related to mineral resources) and transport and communications. The projects were classified as 164 bilateral Arab, 73 multilateral Arab, 91 bilateral Arab–foreign and 99 multilateral Arab–foreign (Sayegh, 1983). Superficially, these figures show good Arab co-operation, but when one analyses in depth the qualities of these joint ventures, many obstacles appear. The major ones are: the geographical limitations of the projects, their joining only with respect to finance, limitations in scope, the fact that many situations are of low efficiency and hence productivity, and finally a rather modest amount of money. As a consequence the expert study, in Sayegh's words, proposed a 'new framework' for complementarity whose centrepiece is the Strategy for Joint Arab Economic Action. This calls for effective vertical and horizontal inter-Arab co-operation. Without going into detail, it is worth mentioning that three of the Strategy's significant suggestions were thrown out by the Arab Summit Meeting: $15 billion was requested over five years, but only $5 billion was approved over 10 years; the money was to go to six less well developed Arab countries instead of to joint programmes; and a council of prime ministers was not appointed (as suggested) to be the appropriate authority to observe and control strategic activities.

If the Arab authorities do not implement, *now*, this and other studies on interaction and mutuality, the future will inevitably hold high pressure. Their peoples, their resources, their economies, and the whole world are pushing towards more sharing and co-operation, so *political decisions can delay but cannot change the natural trend.*

Policies: The International Scale

The industrial, technical and communications revolutions of the nineteenth and twentieth centuries increased the demand for minerals, with a consequent growth in the scale of production and international trade. Further on into this century, the mineral industry became highly monopolistic. The distribution of the benefits of mineral exploitation began to be questioned in countries where mining companies were almost universally foreign-owned and contributions went largely to the parent industrialized countries' markets. These concerns have been surfacing in many developing countries, and reflect the significance of the international mineral resource framework that is forming.

Uytenbogaardt (1977), investigating mineral development policies, stated that among the dangers our finite world is facing are the consequences of an exponential growth of consumption of mineral raw materials. On the other hand, Haglund (1983) has claimed that over the past decade the major industrialized members of the Western economic and military community have intermittently expressed concern about the security of their raw material supply. The policy-makers are acutely aware that minerals constitute an important element in the military potential of nations. There has been a heightened appreciation of the relationship between assured mineral supply at stable prices and the overall functioning of the international economy. Countries must understand the intricacies of this situation and be prepared for whatever might happen in the future.

The mineral interrelationships of the Arab countries with the outside world, which show an almost blind reliance on the non-Arab market were described earlier. It seems that Arab international mineral policies are quite open, but in one direction only. The real benefiting party is the advanced country; the Arab country gets little benefit, and this is often conditional. The problem is not the absence of an international policy, but the presence of one that is foreign-controlled. If the Arab World had a unified policy approach to trading of its mineral wealth on an equal basis with the advanced countries, it would not be standing where it is now. Individual Arab countries are waiting in line for conditions to be imposed upon them, or have no choice but to accept what the advanced countries are offering. Valuable raw mineral matter is traded for a minimal amount of money, valuable partly processed metals or industrial minerals are exchanged for some technology or consumer products of limited worth and, worse, foreign international mineral giants (MNCs or TNCs) are allowed to run Arab mineral activities or processes or operations. Worst of all is when TNCs suck out Arab resources without leaving a proper Arab base that can advance on its own; those TNCs make sure that the Arab country will continuously need their 'good' services.

Were it not for the somehow coherent stand that the OPEC and OAPEC authorities took in the 1970s and early 1980s, they would not have achieved the influencing power that they enjoyed. By the same analysis, it is OPEC's mid-1980s fragmentation in opinion on oil prices and how to handle the market that has lessened influence in the oil world. This fragmented relationship was expressed by Dr Oteiba, the oil minister of the United Arab Emirates, in an Arabic poem (published by the news agency Reuter in daily newspapers, 8 December 1985) implying distrust among member states, and that the unilateral policies they are currently following would lead to further reductions in oil prices and further weakening of OPEC on the world stage. It should be noted that as a group the Arab nations form a sizeable economic block. In 1985 their combined GNP reached $350 billion, and they play an important role in international trade, with aggregate imports of approximately $120 billion and aggregate exports of $160 billion. It is imperative that they identify economic challenges and explore available policy options to enhance their world standing.

On an international scale, Arab involvements in foreign mineral activities (especially funds for developing non-Arab resources) can be seen as being bimodal. One mode is oriented towards the advanced countries and the other towards the developing countries. The former is somewhat negative for full Arab benefit, while the latter is more positive. Let it be very clear that this statement is not made from bias against the advanced countries, because they do spread their technology and help, although politically selectively. It remains certain that no Arab country can ignore the everlasting dilemma of the Middle East, namely Palestine, and how the advanced countries manipulate this issue. Among the developing countries, relationships with equal benefits are dominant. The two modes often clash with each other, although in trade and world policies clashes and relationships occur under diplomatic cover. The 'help' that advanced countries give to the Arabs is always conditional; it always contains some technological colonization; it always comes with some form of political pressure; it always ends by depleting Arab resources. This can easily be discerned from the type of 'co-operation' that contains joint projects or contracts. Many examples can be cited: phosphate fertilizers being made available to developing countries at reasonable prices through MNC agencies run by advanced economies; financial help from the European Community for mineral resource projects being given only to 'stable' developing countries (one wonders what world policies are making the others unstable?); barite from India being traded to the Arab oil countries, with inevitable interference from the USA in supplying the needed machinery; completion of the geological exploration of northern Yemen by the British company Hunting excluding Arab expertise; the Abu Z'abel Egyptian phosphate mine being evaluated by the British company Saltrust with other European companies; the Danish company F.L. Smith's contract with Algeria to establish a cement factory when the cement industry is among those long-established in several Arab countries, which could have joined the contract; the difficulty of marketing the Jordanian Dead Sea chemical salts and the ease of exporting them from the Israeli-occupied side of the Dead Sea; and many other examples of this ill-oriented interaction (*Industrial Minerals*, 1978, 1979, 1980; Corm, 1983; Ibrahim, 1983).

Arab relations with developing countries are based on equal treatment, have easy and quite acceptable conditions, and are non-manipulative and mutually beneficial (Al-Mshat, 1983; Farid, 1982). Yet they are still not widespread, are typically financial-oriented, and are largely bilateral

(Kuwait Fund, Saudi Fund, Iraqi Fund, etc.), although in some cases they are implemented by organizations that are multilateral (Arab share in the World Bank, UN Development Programme, FAO, etc.). This co-operation is facilitated by the many Arab organizations whose purpose is to activate Arab money through help for and investment in needy developing nations. Examples include OAPEC, Arab Fund for Economic and Social Development, Arab Bank for Economic Development in Africa (ABEDA), Islamic Bank for Development and others. It should be noted that co-operation is increasing. The Arab help to Africa accounted for only 4.1 per cent of their total help in 1973, jumped to 13.8 per cent in 1974 and was 27 per cent in 1975 (ABEDA, 1982). Arab help to Africa from 1973 to 1978 was split among the different sectors as follows: agriculture 10.4. per cent, primary industries 7.2 per cent, transformative industries 3.6 per cent, energy 7.6 per cent, construction industry 4.4 per cent, tourism and trade 1.4 per cent, transport 14.7 per cent, financial institutes 6.2 per cent, social services 8.7 per cent, balance of payments 26.2 per cent, technical co-operation 1.5 per cent and help for projects 8.1 per cent. It is obvious from these figures that direct mineral resource involvement is minimal. This might be because neither donor nor recipient has enough technology in the mineral resources domain. This is true for, say, Saudi Arabia giving help to a developing country. But one can look at an alternative: countries like Egypt, Morocco or Iraq can give technical mineral-oriented help, and the needed financial assistance can be supplied by an oil-rich country. Would this work? It would, but only if Arab decision-making authorities realize what proper co-operation is all about and, more significantly, if these authorities direct their policies first to co-operation among themselves and then to co-operation with others. The number one priority is to strengthen and rationalize those pan-Arab institutions that already exist, notably the Organization of Arab Mineral Resources (OAMR), the Arab Mining Company (AMC) and their subsidiaries. These are devoted to investing Arab capital, know how and resources in Arab lands and Arab people, and then, in a decreasing order of priority, in developing countries and in advanced countries.

Another approach is to depend on the varied and excellent services that the United Nations Development Programme (UNDP) can offer (Carman, 1977). Still another alternative is to shift reliance away from the TNCs (which are privately controlled) to governments that have proved themselves among the developing nations, e.g. Greece, Turkey, India, Brazil or some South-East Asian countries, or to governments that do not try to maintain hegemony-type or dominance-type relations with developing countries.

In a very realistic tone, reflecting the needs of this and the coming centuries, the European Community stated clearly that the underdevelopment of Third World countries can be removed only by co-operation (EC, 1987). The principal causes for this underdevelopment include: limited or underexploited natural resources (the raw materials in which Africa is potentially so rich are largely unexploited, with only $50 million of the $1500 million invested annually in mineral prospecting worldwide being spent there—Sudan is a typical case); and the structure of the world economy, which operates under a system created by and for the rich. The statement ends by saying that the EC must cope with these problems and develop industrial co-operation that will serve the interests of all concerned, as the development of the Third World is vital to all.

ADMINISTRATIVE FACTORS

The title of the paper given by El-Fathaly and Chackrian (1983) at Georgetown University's Centre for Contemporary Arab Studies Sixth Annual Symposium was 'Administration, the forgotten issue in Arab development'. There could not be a more accurate title, and it is self-explanatory. The authors argued that this problem was widespread in the Arab World and, because of differences among the countries, the solution was not simple. They described administrative problems, some related to colonial times and Arab traditions, others more basically related to a lack of information and databases important for planning and to the socio-political failings of Arab regimes, e.g. political interference, poor training and education, poor public participation, etc. (El-Fathaly and Chackrian, 1983), many of which have been mentioned earlier.

The main aspects of the administrative problems that affect mineral resource development are threefold: (i) proper management of the resources and all activities that go along with their utilization for the welfare of society is linked to (ii) management, which requires planning, which in turn imposes (iii) the requirement for institutes to prepare everything. This section will concentrate on these three aspects as they relate to mineral resource administration, or as they pave the way for the inter-Arab mineral administration that is needed for future development.

Legislature and Management of Mineral Resources

The first issue is the kind of regime in the country. It has been mentioned that Arab countries are quite different in the nature of their regimes and the control they exercise over their mineral resources. Some are quite liberal towards private enterprise (local or foreign), others apply full nationalization restrictions, and others are selective. Many Arab control directives are dictated by decrees and regulations that form a basic part of the laws and constitution of a country. This is very significant to mineral resource operations from start to finish. Afyeh and Mansour (1977) listed several laws concerning mineral resource activities in Saudi Arabia, Egypt, Sudan, Jordan, Tunisia and Morocco. They compared issues related to: classification of mineral matter as surficial or subsurficial; ownership of the resources; stages of exploration and prospecting; special cases in interpretation and application of these issues; resources in offshore areas; and, of course, who has the right to exploit the mineral deposits under national and financial regulations.

The ECWA (1977) report on the development of mineral resources included a briefing on mining legislation in Jordan, Iraq, Saudi Arabia, the Gulf states, Syria, Lebanon and the two Yemens. The report indicated that these countries still need to improve the functional authorities that administer their resources. It added that mining legislation needs improvement, especially as it relates to development of these resources. In a more specialized study, ECWA (1981) detailed the national mining codes of the region and gave a summary of related legislative and managerial aspects. It showed that some countries do not have provisions for proper minerial classification; some have loopholes in dealing with special cases of mine operation; exploitation of the continental shelf is not well advanced; in many instances government revenues are applied by *ad hoc* agreements; many of the mining provisions date back to colonial times and, if amended, have become more confusing; in most cases the bureaucratic process is a major obstacle; environmental impact and employment policies are passive; and, most importantly, the Arab mining codes do not provide for avenues of Arab co-operation, except in a very few countries, which undermines operations of any scale in border zones and/or of a bilateral nature.

There is no doubt that the management of mineral resources is directly affected by the policies practised in the Arab countries. Related policies have been discussed previously. The obvious drawback in Arab mineral resource management is a missing link between the scientific and technical community and the decision-making circles in government. Efficient management requires the gathering of data, selection of significant elements, interpretation and analysis, and then communication of the results to the decision-makers on an accurate, timely and relevant basis. This process must be carried out by a capable scientific/technical community, which is still not well developed in many Arab countries, although it is available to a considerable extent in others. Again this calls for co-operation among the Arab countries, so that they can benefit from each other's experience. This does not mean that industry should be excluded. It is often industry that is the supplier of data for analysis, as it is the prime transformer of raw material to consumable products. This is dimly seen in the Arab World because information exchange and communication at governmental level is not very strong. Exchange and communication should be facilitated by Arab Chambers of Commerce and Industry.

In the Arab World key positions are often occupied by personnel loyal to the regime, irrespective of their specialization. These positions must be occupied by mineral resource oriented, technically knowledgeable people who know how to manage mineral operations. For mineral resource evaluation to play a full part in the social improvement framework there must be full assessment at national and regional levels. Governmental involvement in mineral science and technology includes both basic and applied science (CRS, 1982).

Even if technically and scientifically capable Arab personnel were holding managerial positions in mineral operations, there would still not be full communication. It should be noted that high-ranking Arabs often have a superiority complex, which hinders vertical integration within the mineral resource operation structure. Integration is essential in all mineral activities. It is needed in the field, the research centre, the factory and the market. Economic shifts and foreign competition have created an international minerals industry and an international marketplace that requires communication. Communication is not a separate management skill; it cannot substitute for planning, decision-making, organizing, controlling and motivating skills. Rather, communication is a vehicle for all these management processes (Hautala and Hoskins, 1985). Management shortcomings are many and varied, and together with a poor market and poor government and employee relations, involve poor planning.

Planning Mineral Activities

A direct input of proper management is the planning of all activities, which often involves a complicated network of functions, processes and perspectives. Advanced mining firms usually have some form of corporate planning process. Unfortunately, the available evidence in the Arab World suggests that the process is either poorly implemented or non-existent. Every mineral resource company or government activity needs and must have access to information in order to plan ahead. The mining industry has been undermined by the rash of unexpected economic and political developments and the accompanying market volatility that has marked the past decade. In the Arab countries, mineral operations are often frustrated by their inability to find the right information, with inevitable consequential poor planning. The approaches adopted may be simplistic and misleading, such as using straight extrapolations of prevailing market growth rates to plan future capacity.

In its 1981 working paper on a strategy for Arab mineral resources development within a framework of economic coherence, AOMR detailed a two-stage 10 year plan for achieving this goal. The first stage, from 1981 to 1985, consisted of a unified Arab geological survey and mineral prospecting programme using remote sensing to present preliminary studies on the feasibility of selective mineral industrialization, to decide on economic priorities for implementing projects, to look for optimum mineral economic links with Third World countries and to start executing the projects with material requirements that could be fulfilled. The second stage, 1986–90, consisted of surveying programmes with the emphasis on mineral resources, upgrading and expanding existing Arab mining projects following the feasibility studies and the continuation of detailed studies on Arab mineral resources for development. It is clear that this has not taken place, and especially not in a co-operative manner. Moreover, there are no signs that it will take place in the near future. Yet again the Arabs may not lack technically and scientifically oriented organizations to do the planning, but they definitely lack the political will for full co-operation.

A vital part of the planning process for mineral resources is shared by the government and the private enterprise. The double-sided activity often creates a conflict of interests between the two sectors. Every country has had, at some time, to face up to this issue, with all its political, social and economic implications. This was the topic of a symposium arranged by the Association of Geoscientists for International Development (AGID, 1978). It has been seen that the Arab countries have diverse political systems, and in most there is no co-ordination between the two sectors. This leads to negative results, in that human and material efforts are wasted in duplication. Mineral development operations, because of their inherently intensive nature, can only prosper within an overall coherent national and regional plan.

Successive Arab governments have planned unilaterally and the resulting strategies have sometimes conflicted with those of other countries, or even those of the same country! Even when a plan has been made, Arab regimes have almost always ignored it. Sayegh (1985) studied the different attitudes of Arab authorities to planning in development. Some have almost no planning because of the power of the private sector (e.g. Lebanon), some are slow and hesitant (e.g. Morocco), some are 'jumpy' (e.g. Sudan), while others have a good and consistent approach (e.g. Tunisia). Countries putting strong efforts in this direction are Egypt and Libya (though with political problems), Jordan and Saudi Arabia, (though with social problems). A positive, consistent and deep planning attitude has been adopted by Syria, Algeria, Iraq and Kuwait. Although some of these countries have shown consistent planning attitudes for decades, they are still lagging. The uniting of efforts in a region towards a coherent plan is the first requirement for this plan to be executed in a logical and meaningful manner.

Arab Institutes of Mineral Resources

In an earlier discussion of institutes, I emphasized training and the creation of a solid human technical base for mineral exploitation. The institutes are also important in the administrative process, especially the geological surveys and specialized mineral resource divisions or companies. Table 1 showed that all Arab regions have geological surveys of some sort and Table 2 indicated the long time for which most Arab countries have been involved with mineral operations. These activities have some quantity but not much quality. The institutes involved need upgrading, and this requires material and human know-how. Since very few Arab countries have both of these aspects, any unilateral upgrading will always fail to some extent. It would be easy, given the political decision, for two or more geological institutes in neighbouring countries to form a stronger and a wider-spectrum service institute. This could remove many problems, such as the lack of staff or money, increase exposure to new techniques and methodologies, and, most importantly, reduce the reliance, sometimes blind, on foreign personnel.

There has been some co-operation, in which certain institutes have used the consultative services of other enterprises or experts, but this has been limited and often on a personal basis (Afyeh and Mansour, 1977; Jabr, 1980; Khawlie, 1983a; Muharram, 1984). Furthermore, the few co-operative steps taken in the mineral resource sector by government institutes have been often disrupted, as between Algeria and Yemen, Egypt and Libya, Morocco and Tunisia, Saudi Arabia and Sudan, Syria and Jordan, Syria and Iraq.

There was a qualitative breakthrough with the establishment of regional Arab institutes, such as the Arab Mining Company and the AOMR, and one would have expected co-operation among the Arabs to increase rapidly. However, in spite of great human and material efforts, it is obvious now that these institutes have not had a chance, let alone priority treatment. It seems either that there is mistrust between Arab mineral authorities and the established institutes, or that the decision to depend fully on Arab expertise simply has not been made.

Details of Arab institutes related to mineral resource activities (i.e. government geological surveys and private sector enterprises) are given by Afyeh and Mansour (1977), ECWA (1977, 1981), Jabr (1980) and various national papers presented at Arab conferences on mineral resources. It is noteworthy that these surveys (which are better termed 'geological bodies', as there is no uniformity in nomenclature) are controlled by a variety of ministries: the ministry of oil and energy resources, the ministry of industry and energy, the ministry of commerce and industry, the ministry of oil and mineral resources, the ministry of finance, the ministry of industry and mining, the ministry of resources and energy, the ministry of development—probably only in Somalia is there a specialized ministry of mineral resources and water. In some countries, such as Jordan and Yemen, they are under direct control of the prime minister. This diversity in perspective and direction could make it difficult for these bureaucracies to work with each other. The above references give figures on the types of specialized services (mapping, geochemical exploration, geophysics, etc.) and an idea about Arab specialists. They all agree that the Arab World still has a long way to go in this respect.

In addition to these institutes, and the two mining training institutes in Morocco and Jordan, there are a number of specialized higher education colleges and universities or applied industrial institutes, such as the University of Petroleum and Minerals and the Centre for Applied Geology, both in Saudi Arabia, the Colleges of Petroleum and Mining in Libya and Egypt, and several specialized geology and engineering departments. It is likely that there is duplication with the associated waste. Some of the institutes are very well equipped in laboratory facilities and advanced computer applications, while others are lucky in having highly experienced personnel. It is rare to find both together, and where they are they lack sufficient finance. This pattern is typical of the underdeveloped or slightly developed countries. No doubt if the Arab authorities complemented each other's efforts they could be much more positively productive.

At the recommendation of the committee of the Fourth Arab Mineral Conference, held in Amman in 1981, the AOMR did a study to enhance and upgrade the exploration and prospecting of Arab mineral resources. The detailed study made well-planned and documented procedural suggestions on technical and financial aspects, and indicated that because of impaired functionality of the Arab geological/mineral related institutes, the evaluation of Arab mineral resources is still inadequate. The reasons for this are many and varied, including finance, infrastructure, trained personnel and the poor state of science and technology. Although the study was made available in 1981, the same problems were reflected by the AOMR main working paper in 1985. This clearly points to something very worrying in the Arab countries: they form co-operative pan-Arab organizations (the AOMR is only one example), they ask these organizations to carry out needed investigations, the organizations come up with high-calibre results, and the train stops there! There is no execution, and there seems to be no intention to get involved with collective projects.

Finally, it should be noted that the Arab mineral resource institutes must share the blame. Many are well structured but not necessarily well managed. One sign of their weakness is that they do not try to influence the decision-makers. Nor do they try to share their work with the general public. The institutes stand as secluded islands within their own societies, so it is not surprising that they do not extend their help or look for effective co-operation with other countries.

FINANCIAL FACTORS

During the recent past, the international economic environment has been unsettled, and the Arab countries have sought to adjust to significant fluctuations in prices, slower growth, increasing restrictions on international trade and difficult financial markets.

Although Egypt benefited from a growth in revenue from petroleum exports, the impact of the international recession has widened the country's deficits. Some corrective measures have been taken, but they have fallen short of a fuller exploitation and mobilization of domestic resources. Algeria's economic management policies have cushioned the destabilizing impact. Among other measures, the authorities implemented a programme of governmental reorganization, including regional and public sector decentralization, financial restructuring and (very important for the minerals domain) incentives to small-scale activities in the private sector. These are only two examples of Arab countries facing their financial positions, which have different inputs and outputs, and hence different effects on their development.

There is no doubt that the economic base depends heavily on resources. These include natural resources, the availability of technically trained manpower and, crucially, the availability of finance for investment within a well-defined economic interactive policy. This section will deal with three issues related to financial factors in Arab mineral resouces: investment in mineral-oriented projects; the industrialization of raw mineral matter; and the countries' trade positions in the market.

Investments in Mineral Activities

One of the most basic economic determinants in the development process is the availability of finance for investment. It is equally true that the economic feasibility of a project directly influences the decision-maker to give or not to give the green light. Mineral resources can play a positive or a negative role in production diversity and efficiency in the economy. This depends on the resources' availability and value, and the ease of prospecting and processing. The mineral resource base is not static or independent. It is a reflection of what is available in the country in terms of financial capital, technology, research, markets, industry, etc., and these are not equally distributed in the Arab countries, at least not yet.

The answer to the obvious question, of whether the proper finance for mineral resource development is available in the Arab World, is no, and this might shock many observers. In the 1970s only Saudi Arabia, Kuwait and Libya were in a sufficiently good financial position to hold enough hard currency to secure their imports, capital products and technical services (Sayegh, 1985). It is not enough to have the needed finance, it must also be absorbed efficiently in the development process. The past decade in the Arab World, as a whole, has been one of intense scarcity of finance, although a very few well studied projects have been undertaken. These projects were in oil-rich countries. Other projects, and other Arab countries, have suffered badly.

In another study of Arab financial resources, Abed (1983) suggests that the manner in which these resources have been deployed has produced limited rewards. The rapid accumulation of external assets in the 1970s, caused by the Arab oil countries' limited absorptive capacities, was a short-term phenomenon. Abed indicates that these countries generated surpluses largely to accommodate an expanding world demand for energy. He concludes that the political fragmentation and institutional under-development of the Arab World reflects the absence of a coherent financial deployment strategy, and prevents the formation of a coher-ent view of Arab strategic interests, which the resources are presumed to serve.

Al-Sadik (1985) referred to this situation as the 'income illusion of petroleum exports'. He said that the mainstay of the oil countries is a dwindling resource, which leads to a confusion of the two concepts of wealth, which is a stock concept, and income, which is a flow concept. The income illusion shows up in: the overestimation of national income level, national savings and accumulation of national wealth; the underestima-tion of domestic absorption and foreign aid; and distortion of the current account position and the contributions of the different sectors to national income.

Against this background one can evaluate the financing of Arab mineral activities without bias from the 'income illusion'. The papers submitted to the Fourth Arab Mineral Conference (summarized by Khawlie, 1983a) give a good picture of Arab mineral investments. The projects were concerned with deposits of major resources, including phosphates in all the regions except the Arab Peninsula, copper in all except the Nile Valley, and iron, lead, zinc and mercury in the Arab West. The investments, covering a range of two to ten years, i.e. up to 1990, totalled approx-imately $2.5 billion in the Arab West (mainly Algeria and Libya), $1.0 billion in the Nile Valley (mainly Egypt), $5.2 billion in the Fertile Crescent (mainly Iraq) and $2.0 billion in the Arab Peninsula (mainly Saudi Arabia). These covered all aspects of mineral development. Unfortunately, the Fifth Arab Mineral Conference held four years later did not suggest that the plans had been implemented. With a few exceptions, there was poor productivity and exploitation (much worse than expected). Some minor projects were completed, and small steps were taken in the major projects. Nevertheless, the Arab mineral authorities had at least made a more defined and better selection of those mineral projects that proved more economically feasible. If inter- and intra-regional co-operation had been widespread, the productivity and efficiency of the Arab mineral scene would have been transformed.

It is remarkable that the $10.2 billion invested would have revolutionized the Arab mineral scene if it had been applied as planned. Of course, it was not. Compared to estimates of the need for mining finance in the same period in the whole non-socialist world—$15 billion per year (Radetzki and Zorn, 1979)—the Arab investment is quite considerable, especially because 40–50 per cent of the $15 billion would be from the developing countries. But Arab financial strength would be much more effective if a coherent collective plan was followed, channelling the needed liquidity into the mineral sector. The total Arab GDP in 1982 amounted to $392 billion, so these investments would have constituted a tiny fraction of Arab finance. The extractive industry had a growth rate of only 18.7 per cent from the mid-1970s to the early 1980s (AOMR, 1985). The Fifth Arab Mineral Conference concluded its section on 'the economic position of Arab mineral resources' by emphasizing the need to lessen the huge gap between Arab mineral potential and its actual economic development, adding that this could be achieved by executing the strategies and policies *already recommended* in previous conferences (AOMR, 1985).

The financial problem is not the absence of investments (if a coherent plan is adopted) but the absence of proper financial channelling of both internal finance and external aid or investment. There are several factors to be considered in this respect. First is the national (and regional if complementarity is the goal) levels of income, modernization and progress in the mineral sector. Second is the assumption(s) about mineral activity growth, the capital required and how far it can be secured internally and supplemented externally. The third consideration is the absorptive capacity of the social structure, in the mineral resource sector on the one hand, and the whole nation or region on the other hand. Financial investments need detailed investigations of economic, technical, social and managerial aspects. There is an obvious selectivity in investment that is decided by a priority plan of the intended mineral resource projects. This is particularly important because Arab potential financers are few in number, i.e. the oil-rich exporting countries; these expect to have a say on priorities. It also seems that the Arab World is entering an economic crisis that will last through the mid-1990s, with growth not being restored before the year 2000. The governor of the central bank of the UAE declared in January 1986 that the Arab countries should aim primarily to rid themselves of their external debts and foreign economic dependence.

The worldwide financial state of the mineral sector does not look very healthy. Raw mineral output, which supports the production of primary metals, non-metal mineral products, chemical and other allied products and industries, did not benefit from the partial recovery of the economy (O'Neil, 1985). Strauss (1983) stated that the economic recession of 1980 was prolonged and severe but especially so in its effect on the mineral industries. The capital costs of new mineral projects or of expansion of existing facilities have escalated at an alarming rate since the early 1970s. The real cost of establishing a new mining and processing operation, measured in constant dollars per annual tonne, has increased over five times from the early 1970s to the early 1980s (Radetzki and Zorn, 1979). A recent research report by Intermet Publications analysed what happened to 67 mineral commodities in the 30 years since 1950 (*Industrial Minerals*, 1983) in terms of prices, costs, production levels and markets. The report concluded that the present recession has deep roots and dates back at least 10 years. Base metals such as copper, lead and zinc were sold in the 1970s and recent years at prices 10–30 per cent lower than historically, if expressed in constant currency values. Things were much better in the non-metallic sector, and the main reason is more balanced production and demand.

Investments clearly play a complex role in mineral development. Attention should be given, at all stages of mineral activity, to preliminary market factors (transportation, specifications, supply and demand, and prices) and these should be correlated with technical input. If the necessary studies are not done, the priority task of identifying specific mineral projects, and then identifying potential investors, will not be achieved. As an example of what can be done, the Turks have updated their laws, speeded up their decision-making process and most importantly kept bureaucratic intervention to a minimum, and have attracted foreign investment. The future trend in the mineral sector is towards exploitation of the more available lower-grade deposits, large and complex plants, high-technology products and diversification. The goal must be continued growth, and throughout this growth process development can be achieved by awareness of the market, of the basic principles of the demand, of specifications, of spheres of influence and of the price.

A review of recent exploration costs in the Third World by the UN Revolving Fund (the AOMR has recommended that a similar fund for Arab mineral resources be established under the auspices of AOMR, which would replace the many Arab funds, not one of which is specifically oriented towards mineral resources) shows that between $2 and $3 million per project is required to finance an intensive minerals exploration programme. The annual project budget for such a programme is between $600 000 and $800 000, with manpower accounting for around 50 per cent of this (AGID, 1985). Such a financial burden is quite heavy for a single Arab country. Some could not afford it and some could but are not fully capable of the undertaking (because of the oil income illusion). It is inevitable that they will have to co-operate in a co-ordinated, coherent plan that takes economic, financial and intangible factors into consideration. The economic analysis evaluates the relative economic merits of investments from a profitability viewpoint, based on discounted cash flow or a similar analysis of projected revenues and costs. The financial analysis evaluates where and how the funds for proposed investments will be obtained (Stermole, 1983). The intangible analysis considers economically unquantifiable factors affecting investments, such as legal aspects, public opinion and other political and environmental impacts. The processing and transformation of minerals into an effective industry must be considered for the input of the required finance.

INDUSTRIAL FACTORS

It is about time that the Arab countries realize that the huge revenues they have accumulated, notably from oil exports and mainly not through national productivity, will not be of value unless properly channelled internally for the development of the various sectors, industry being the prominent one. Industry diversifies the economy, trains and educates the labour force, creates sources of income and is a carrier of technology. These points were raised by Kubursi (1985), who discussed the industrialization of the Gulf Co-operation Council and stressed that the end of the oil era was in sight. He concluded, after weighing alternatives and considering various networks, that industrialization among the Arabs must be co-operative and comprehensive.

The Arab authorities have realized this, or at least they have been notified about it (many times), and the Eleventh Summit Meeting held in Amman in 1980 adopted a strategy along these lines. The strategy stresses the realization of industrial development, concentrating on basic and engineering industries. These require the exploitation of raw mineral matter, thus making mineral resources the central issue of Arab industrialization. In spite of this and the fact that industrializing mineral resources raises the manufacture value added (MVA), the mineral industry in the Arab World does not exceed 25 per cent of the local mineral productivity, with the rest being exported as raw material. Why?

Many researchers trace the situation back to pre-independence times, and others stress that the Arabs have been late in joining the industrial trend, doing so in a world of uncertainty and monopoly. What is crystal clear is that the Arabs need uniformity, or a coherent attitude to push for a coherent strategic development plan. However, one should not put all the blame on Arab authorities. The world industrial map is changing, as detailed in UNIDO's industrial development survey, the major background document of the Fourth General Conference held in 1984. The major reasons are the new economic constraints that have emerged. The developing countries had 11 per cent of the world's MVA and 66 per cent of its population in 1982 and, if current trends continue, their MVA will be only 14.9 per cent while they will account for 72 per cent of the world's population by the year 2000 (UNIDO, 1983).

The UNIDO study grouped world countries according to various economic criteria. First are those that are developed, with high incomes, followed by those with less development. The Arab countries' positions are shown in Table 13 (modified from UNIDO). The spectrum of Arab MVA growth rates can be compared to the average MVA growth rates of different groupings of countries: centrally planned economies 7.98; industrialized economies 3.2; large developing countries 7.19; small developing countries with modest resources 4.64 (this group includes most of the Arab non-oil countries); and small developing countries with ample resources 6.34 (most of the Arab oil countries). If the two extreme cases (Sudan and Libya) are eliminated, the Arab countries' MVA is only just acceptable overall, because the average Arab MVA, about 6.0, is lower than that of many developing Third World countries that do not have the natural resources that the Arabs have. This implies that Arab industry is deficient in the variety of manufactured products produced, lagging in its involvement in the social structure and welfare and weak in exploitation of the Arab natural resources. So the mineral resource sector, as a transformative industry, is terribly handicapped.

The Arab extractive industry (mostly oil) grew in the 1970s by about 40 per cent while the transformative industry grew by less than 17 per cent. Local demand for industrial products exceeded by three times local industrial productivity, leading to a high reliance on foreign imports. Industrial labour makes up less than one-tenth of the total Arab workforce, and industrial income is about 6.5 per cent of total income. Consumer products account for 50–90 per cent of total Arab industrial production. Most of this industry is located in only six Arab countries: Egypt, Morocco, Algeria, Tunisia, Iraq and Syria (Abdel Nour, 1981). These figures show a marginal role being played by industry.

Many of the developing countries have followed industrial policies that are appropriate to their desired growth and social development, and some have succeeded, even with minimal resources. Malawi, for example, is one of the poorest countries in the world, but it was praised by a World Bank report as a good example of how a small African country with little

Table 13
GROUPING OF ARAB COUNTRIES BY ECONOMIC CRITERIA, AND MVA GROWTH RATES FROM 1970 TO THE EARLY 1980s

Group (all are developing countries)	Region			
	Arab West	*Nile Valley and Western Red Sea*	*Fertile Crescent*	*Arab Peninsula*
Non (or weak) oil countries				
Least developed		Sudan −0.6(−) Djibouti 4 Somalia 5.2(+)		Yemen S. 2.6 Yemen N. 3.3
Lower middle income	Mauritania 4.4			
Middle income		***Egypt 6.4		
Higher middle income	**Morocco 5.6(+) **Tunisia 11.6(+)		*Lebanon 8(+) *Jordan 11.8	
Oil countries				
Increasing reliance on industrial diversification	Libya 24.3(+) **Algeria		**Syria 6.4(+) *Iraq 10.1	Oman 4 UAE 5 S. Arabia 5.1 Qatar 6 Bahrain 6 Kuwait 12.0

Figures are average annual MVA growth rates.
(+) MVA doubled 1960–80, (−) MVA halved 1960–80.
*, **, ***, trend of increasing industrialization.

apparent industrial potential can enjoy a high rate of manufacturing growth while following an agriculture-oriented development strategy. The government has provided protection for infant industries and actively promoted industry through a well defined social welfare policy (Livingstone, 1984).

The drive to industrialization and its consequential development depend primarily on the industrial policy followed. There are three major alternatives: the import substitution, the export-oriented and the integrated industries. An example of the last is the mineral resource industry, which results in the manufacture of several items that go into the production of many different products. It is a 'chain production' type industry (Yamut, 1984), which must be functional in all economic sectors to provide the essential push to the manufacturing domain, leading to progress in the services domain.

Among the first industrial policies adopted by Arab countries was that of import substitution. This led to some industrial progress and the evolution of some administrative and technical ability. But it was short-sighted and was blocked by: the small size of the market; a horizontal expansion and diversification of the consumer industries at the expense of the intermediate and primary industries; a concentration on raw mineral and agricultural export products; and ineffectiveness in production capacity, notably in manpower. This led several Arab countries to turn to export-oriented industries, to establish free zones, to encourage foreign investment and to undertake joint projects with foreign companies, especially the trans-national companies. Again this led to several negative effects, including: increased reliance on the outside world; foreign control of the technological process; uncertainty about economic output because of its links to the world market; a reduction in dependence on local labour; and a failure to create enough local technically trained manpower.

This situation opened the way for inter-Arab mutuality, which surfaced in the early 1980s and was accentuated by the Eleventh Arab Summit. Although a plan was formulated and its first programme (1981–85) should be complete, there is no sign of it, so the same old problems of complementarity continue. The plan's industrial projects would be of the collective type, i.e. an industrial plan within a time interval, consisting of several chained programmes, each containing many related projects distributed among the various Arab countries, and individually executed and administered (Rashīd, 1984). Since the establishment of the Arab Organization for Industrial Development (AOID) within the Arab League, many efforts and studies have tried to encourage and co-ordinate this spirit (Masha'l, 1977). Masha'l emphasized, even in 1977, the necessity of diversifying income sources by exploiting mineral resources and processing iron ore into steel, in a two-stage plan: 1976–85 and 1985–90, with investments of $20–22 billion and $25–50 billion, respectively. Where does this stand now? As mentioned in the previous section on finance, investment has been much lower than expected, and the mineral resources industry is still terribly backward. The most obvious evidence is in the Arab steel industry, which is among the early mineral industries, having existed for almost 40 years. It is still almost completely restricted to the construction sector, and more than 70 per cent of the primary production goes as foreign exports (Dimachkieh, 1984). International markets for steel have severe competition and uncertainty in prices. Moreover, Arab production capacity is about 15–20 per cent of apparent Arab consumption. The conclusion about this long-standing industry must be that there is no integration within or between sectors, even in the same country!

The economic problems of most Arab countries, like those of other developing countries, could be eased, if not completely eliminated, if the Arabs were to explore and exploit to the full their mineral wealth. The Arabs have not been able to utilize their mineral wealth to improve the quality of life of their people. The inadequate MVA and the non-integration of the mineral export industries into national economies leads to a lack of backward and forward linkages.

A comparison of the fifth and sixth Arab conferences on industrial development is interesting. The Arabs saw in the fifth conference, held in Algeria in 1979, that an industrial strategic plan should be implemented, considering industrialization as a necessary condition for total economic and social development. They agreed that it should be based on two premises, one concerned with increasing the welfare of the people, and the other with co-ordinating all efforts towards this goal. This meant creating an industrial sector that integrated the consumption, intermediate and production industries in a continuous complementary chain. Five years later, the sixth conference was held in Damascus. In this conference experts discussed studies concerned with: the status of Arab industrial development; capital industries; iron and steel, chemical fertilizers; petrochemicals; clothes; food industries; and construction industries. The inadequacy of Arab industrial development was noted in several areas, including: co-ordination between the AOID and related industrial facilities, e.g. governments, companies and experts, especially in economic feasibility studies; research and development; technology transfer and its implantation; encouragement of the private sector in inter-Arab joint projects; the updating, upgrading and dissemination of full statistical data on industries; and the training of different levels of manpower, making use of existing institutes. The conference produced a working paper that was presented to the Council of Ministers. This paper categorically pointed out that transformative Arab industries still depend heavily on the import of intermediate products and raw materials as a consequence of the weak, almost absent, *integrated* internal (intra-Arab) and inter-Arab industrial circuits (AOID, 1985). So, what integration has been achieved since the fifth conference? Almost none! The working paper also pointed out the necessity of capitalizing on basic co-operative Arab programmes when choosing joint project(s), so that maximum benefit will prevail. Before looking at these basic programmes, we shall examine the standing of the mineral industry in the Arab World (Figure 3) (data from Dempsey, 1983; US Bureau of Mines, 1983; World Bank, 1984; United Nations, 1984a, b, c, d; Europa International, 1985).

The picture that emerges is one of contrast. Although most, if not all, Arab countries have been stressing industry and mineral industrialization, most fall in the low to moderate industry indicator categories. More important still is that although some countries may show a high industrial

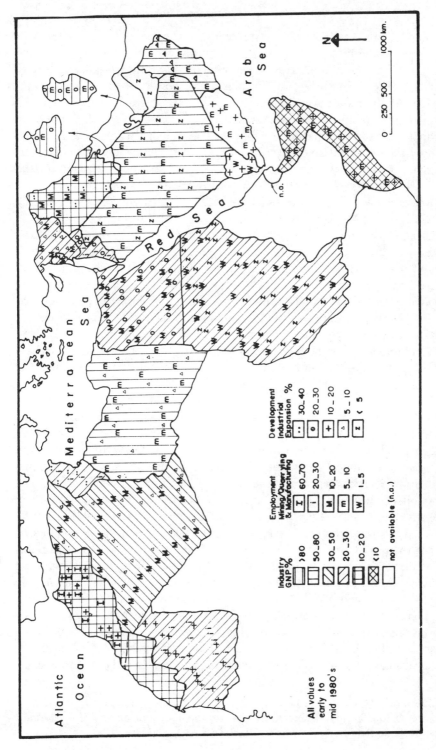

Figure 3. Arab industrial indicators: GNP, mining, quarrying and manufacturing, and industrial expansion.

63

national product, these same countries do not have a high percentage of manpower in mining, quarrying and manufacture, or are not aiming for a large expansion in industrial development. This may reflect not only improper industrial mineral planning, but also shortages in manpower, finance or the required infrastructure. These problems can be eased to a great extent if and when Arab co-operation takes place. This has been indicated before: different Arab countries co-operating together can supply the technical manpower, finance and resources needed to meet the demands of an Arab mineral industry. UN statistics can be used to find the relative degree of industrialization (RDI) (UNIDO, 1982) for some representative Arab countries (Algeria, Tunisia, Sudan, Iraq and Syria). When these RDIs are compared to the RDIs of developing countries in general, Sudan has a much lower figure, Syria shows a random picture, and the others are much better. A notable feature, however, is that all have lower RDIs than the generality of developing countries in iron and steel, and in non-metallics such as clays, pigments, grinders, insulators and refractories. This is less true for the non-ferrous metals, and even less so for metal products excluding machines. The items chosen from the UN data are directly related to mineral matter, i.e. industrial chemicals, other chemicals, petroleum refinery products, pottery, china and earthenware, glass, other non-metallics, iron and steel, non-ferrous metals, and metallic products excluding machinery.

The industrial picture, in terms of these partially processed resources and the indicators given in Figure 3, makes more sense if it is linked to the geographical distribution of the mineral resources so far discovered in the Arab World. Figures 4 and 5 show the distribution of metallic and non-metallic mineral resources in the Arab World to the best approximation from national papers presented to Arab mineral conferences. The first observation is the huge area of land still unexplored, notably in Mauritania, Algeria, Libya, Sudan, Somalia, Iraq and the Arab Peninsula. Most Arab land remains unexplored. The second observation concerns those resources that are widespread in the Arab World, e.g. the phosphates, and that would justify the construction of *regional* processing centres. Another point to be considered is the large steel industry in Egypt, which is quite justifiable because the region is rich with ferrous elements and hence all effort should be made to strengthen it. It is to be hoped that the industrial processing of other mineral resources becomes distributed regionally not only in accordance with availability, but also in association with regional development plans. This should put emphasis on

---

Figure 4. (pp. 66–71) Schematic maps of the distribution of Arab metallic resources. (a) Region A; (b) Region B; (c) Regions C and D.

the inner undeveloped areas, and would require more interaction between the industries of different countries. For example, the steel potential of Mauritania requires the transfer of Arab manpower, advanced technology and finance; the rare metals potential of Saudi Arabia requires the transfer of Arab manpower, technology and infrastructure; the heavy rare-earth sands in Somalia and the Yemen, the uranium and tungsten of Algeria, the micas and fibre minerals of Sudan, and many more, require the establishment of a highly diversified processing network spread over all the Arab countries.

Variably processed (industrialized) Arab mineral resources that are exported in considerable quantities are: fertilizers, crude and manufactured phosphorus, nitrogen, sodium and potassium compounds; construction materials; cement and lime; refractories, ceramics, bricks and other china and earthenware; some smelting, refining, concentrating and alloying of aluminium, copper, lead, zinc, silver, cobalt and nickel; iron and steel universal plates, sheets, tubes, castings, primary forms, pipes, fittings, bars, etc.; glass; salts and brines; abrasives; sulphur products; and precious or semi-precious stones. All other mineral-related industries are still very small, and most of those listed above do not undergo full processing or manufacture. Arab mineral industrialization needs, and can still take, very long steps forward. The programmes of the sixth Arab industrial conference give an idea of whether the trend, *if implemented*, is in the right direction. Table 14 (modified from AOID, 1985) shows only the mineral resource projects that would process Arab raw mineral matter into various industrial products. These comprise seven out of a total of 20 programmes. Investment in them makes up about 22 per cent of total investment. The labour force, however, makes up only 5–10 per cent of the total. Does this reflect a healthy exploitation of minerals? Not really.

Two blocs of countries can be taken as examples of Arab industrial direction and status: the Gulf Co-operation Council (GCC) and the Arab West. The countries of the GCC have realized the danger of depending on one source of income, oil, and have recently started diversifying. However, industry made up only 12.5 per cent of their economic development plans, and was linked to petrochemicals. The contribution of private enterprise was tiny, and mostly in light consumer industries. The productive sector generally showed a varying growth of MVA, further increasing in the oil sector and decreasing in the others. Co-operative, integrated industrial projects remained almost absent (Hashimi, 1984). In a study of the integration of industry in the Arab West, Lehzami (1980) emphasized the ultimate necessity of a mutual industrial policy for developing this region, with special attention paid to the mining and transformative industries. At present they are in chaos. Algeria has chosen heavy industry; Morocco has changed to basic industry; Tunisia

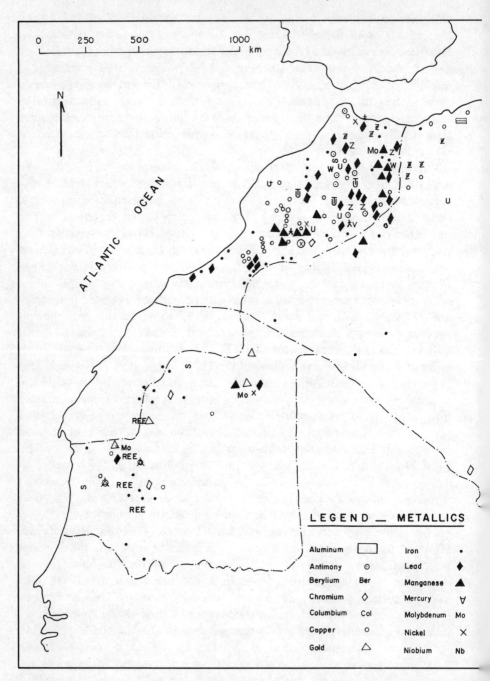

LEGEND — METALLICS

Aluminum	▭	Iron	•	
Antimony	⊙	Lead	◆	
Berylium	Ber	Manganese	▲	
Chromium	◇	Mercury	Ɐ	
Columbium	Col	Molybdenum	Mo	
Copper	○	Nickel	✕	
Gold	△	Niobium	Nb	

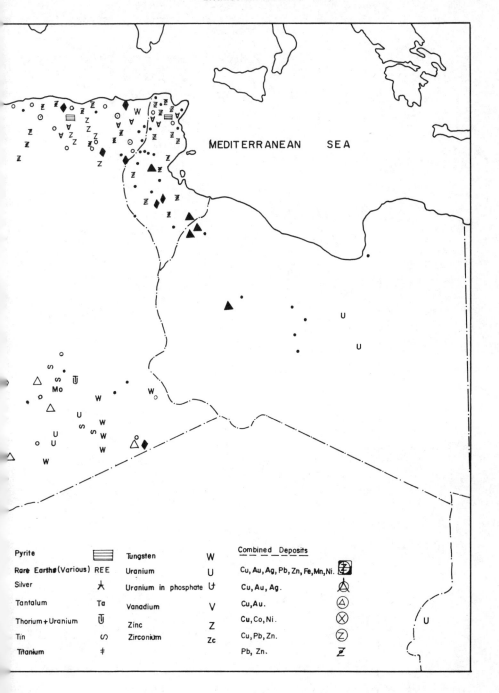

Pyrite	⊟	Tungsten	W	Combined Deposits	
Rare Earths (Various)	REE	Uranium	U	Cu, Au, Ag, Pb, Zn, Fe, Mn, Ni.	⊞
Silver	⅄	Uranium in phosphate	Ʉ	Cu, Au, Ag.	⊼
Tantalum	Ta	Vanadium	V	Cu, Au.	△
Thorium + Uranium	Ū	Zinc	Z	Cu, Co, Ni.	⊗
Tin	S	Zirconium	Zc	Cu, Pb, Zn.	Ⓩ
Titanium	‡			Pb, Zn.	Ƶ

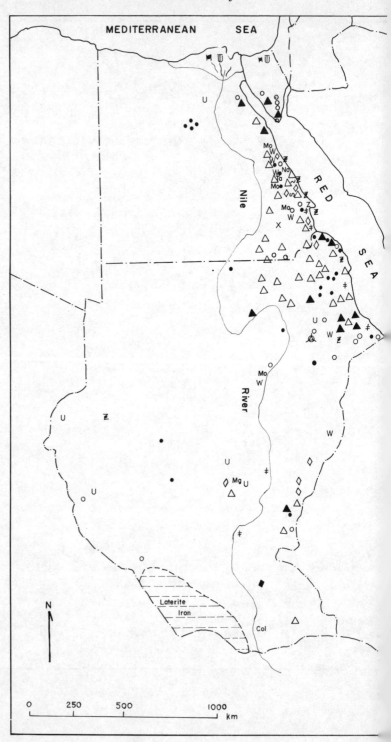

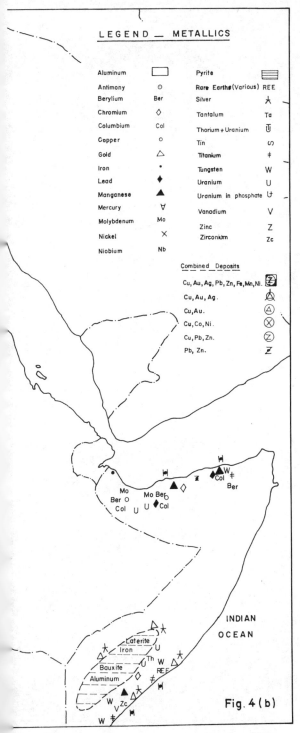

LEGEND — METALLICS

Aluminum	▭	Pyrite	☰
Antimony	⊙	Rare Earths (Various)	REE
Berylium	Ber	Silver	人
Chromium	◇	Tantalum	Ta
Columbium	Col	Thorium + Uranium	ᵁ̄
Copper	o	Tin	∽
Gold	△	Titanium	≠
Iron	•	Tungsten	W
Lead	◆	Uranium	U
Manganese	▲	Uranium in phosphate	ᵾ
Mercury	∀	Vanadium	V
Molybdenum	Mo	Zinc	Z
Nickel	×	Zirconium	Zc
Niobium	Nb		

Combined Deposits

Cu, Au, Ag, Pb, Zn, Fe, Mn, Ni.	⊞
Cu, Au, Ag.	Ⓐ
Cu, Au.	Ⓐ
Cu, Co, Ni.	⊗
Cu, Pb, Zn.	Ⓩ
Pb, Zn.	Z̄

INDIAN OCEAN

Laterite
Iron
Bauxite
Aluminum

Fig. 4 (b)

69

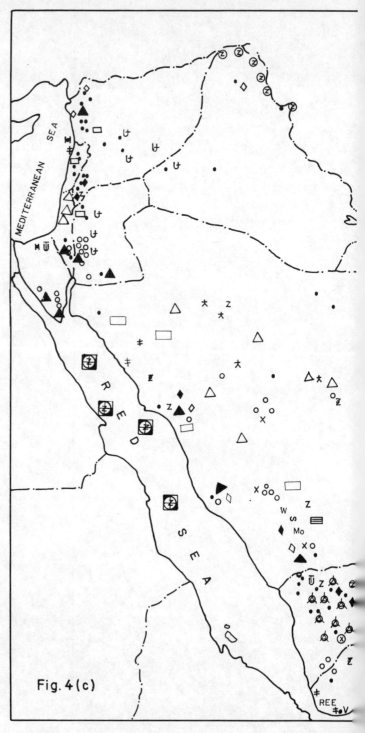

Fig. 4(c)

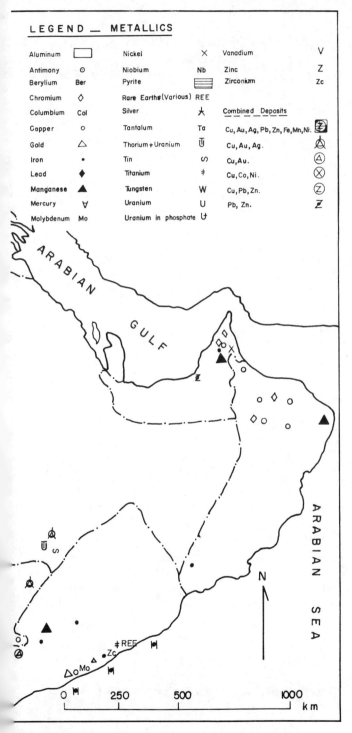

LEGEND — METALLICS

Aluminum	▭	Nickel	×	Vanadium	V	
Antimony	⊙	Niobium	Nb	Zinc	Z	
Berylium	Ber	Pyrite	▤	Zirconium	Zc	
Chromium	◇	Rare Earths (Various)	REE			
Columbium	Col	Silver	人	Combined Deposits		
Copper	○	Tantalum	Ta	Cu,Au,Ag,Pb,Zn,Fe,Mn,Ni.	⊞	
Gold	△	Thorium + Uranium	Ū	Cu,Au,Ag.	⊛	
Iron	•	Tin	∽	Cu,Au.	Ⓐ	
Lead	◆	Titanium	ǂ	Cu,Co,Ni.	⊗	
Manganese	▲	Tungsten	W	Cu,Pb,Zn.	Ⓩ	
Mercury	∀	Uranium	U	Pb, Zn.	Z̄	
Molybdenum	Mo	Uranium in phosphate	Ů			

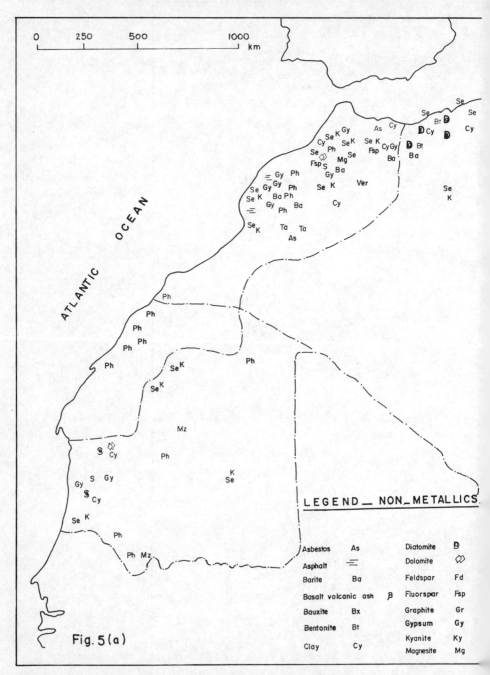

Figure 5. Schematic maps of the distribution of Arab non-metallic resources. (a) Region A; (b) Region B; (c) Regions C and D.

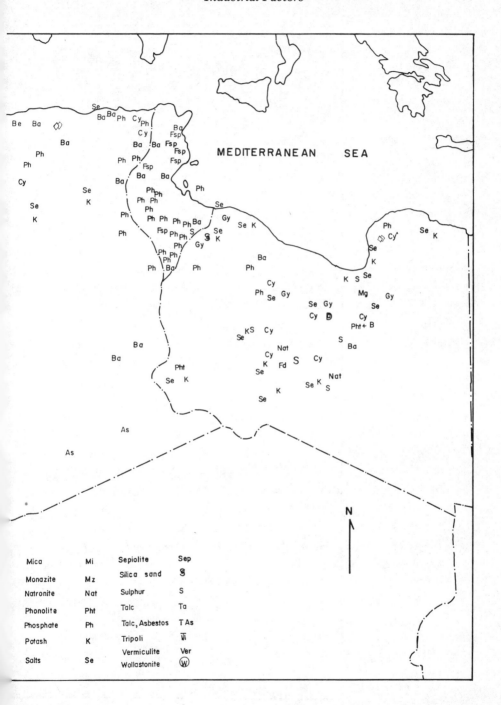

Mica	Mi	Sepiolite	Sep	
Monazite	Mz	Silica sand	S	
Natronite	Nat	Sulphur	S	
Phonolite	Pht	Talc	Ta	
Phosphate	Ph	Talc, Asbestos	T As	
Potash	K	Tripoli	T̄I	
Salts	Se	Vermiculite	Ver	
		Wollastonite	Ⓦ	

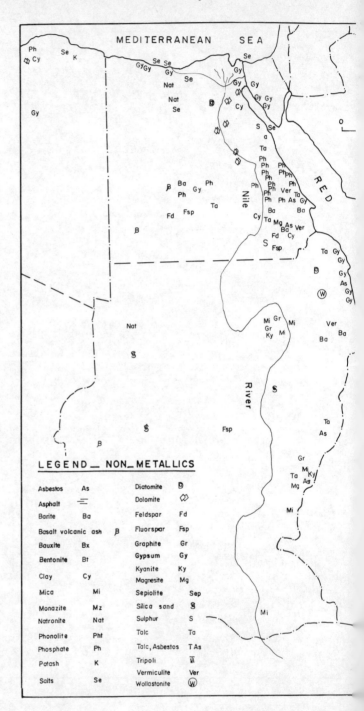

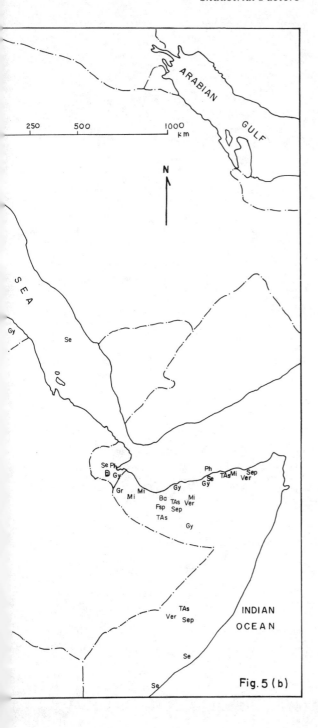

Fig. 5 (b)

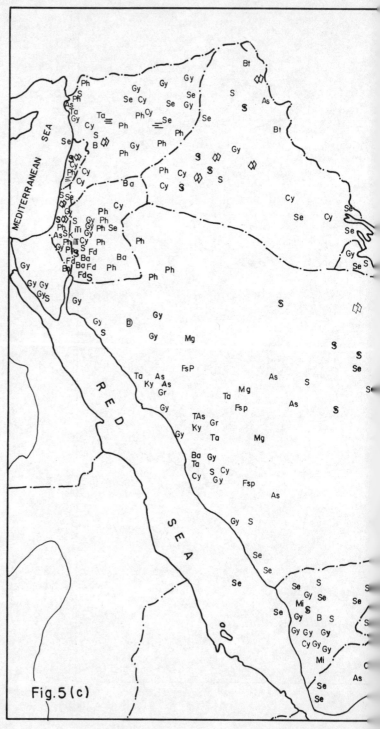

Fig. 5 (c)

LEGEND — NON_METALLICS

Asbestos	As	Dolomite	⟡	Phonolite	Pht
Asphalt	☰	Feldspar	Fd	Phosphate	Ph
Barite	Ba	Fluorspar	Fsp	Potash	K
Basalt volcanic ash	β	Graphite	Gr	Salts	Se
Bauxite	Bx	**Gypsum**	**Gy**	Sepiolite	Sep
Bentonite	Bt	Kyanite	Ky	Silica sand	Ⓢ
Clay	Cy	Magnesite	Mg	Sulphur	S
		Mica	Mi		
Diatomite	Ꭰ			Talc	Ta
		Monazite	Mz	Talc, Asbestos	T As
		Natronite	Nat	Tripoli	Ⅶ
				Vermiculite	Ver
				Wollastonite	Ⓦ

Table 14

ARAB MINERAL RESOURCE PROGRAMMES ANNOUNCED AT THE
SIXTH ARAB INDUSTRIAL CONFERENCE

Programme (locality)	Arab demand (thousand tonnes)	Labour needed	Number of projects
Iron pellets (Mauritania)	6000 (Egypt, Algeria, Libya)	300	1
Graphite rods (Egypt or Libya; later Algeria and Gulf)	96–120 per year	680 per project	2–4
Special steel (Gulf; later Arab West and East)	725; capacity 200 per year; investment $451 million	3475	1–3
Combined fertilizers (Arab West)	Demand not well defined; capacity 1000 per year; investment $40 million	?	1
Concentrated fertilizers	44 000 in 1990, 608 000 in 2000	?	?
Glass sheets (Arab West; later Iraq)	514 in 1990, 1236 in 2000	125 per project	1–2
Refractories (Iraq, Libya)	Demand not well defined; capacity 100 per project; investment $180 million per project	?	2

? = not well defined.

has replaced its mining industry by light industry; Libya has diversified
to a considerable extent; and Mauritania lacks a well defined industrial
policy. Unilateral decisions are obvious. Market distortions have resulted
in more economic problems and an inability to build up the infrastructure.
These market distortions can be classified as technical barriers, market imperfections, domestic policy inconsistencies, lack of commercial
viability and information gaps. The eventual consequence will be the
failure of a potential natural resource processing activity to develop
because of the negative impacts of these distortions on economic and
commercial evaluations of projects (UNIDO, 1979, 1981c). There are six
processing stages of raw mineral matter, and these have rarely been fully
achieved in Arab countries. In increasing order of complexity, they are:
raw material; processed raw material; semi-processed product; first transformation or raw finished product; second transformation or simple
finished product; and the complex finished product.

For any viable project in the minerals industry, available economic
deposits, technical expertise, finance and market accessibility must all be

secured. Arab countries in isolation or following unilateral policies cannot reach strong industrial positions. A co-operative policy among Arab countries is feasible, and is plausible, because the necessary organizations do exist and their joint efforts could overcome most of the disadvantages so far faced by all Arab countries.

It remains for this industry to contribute a decent share to the economies of the country, the region and the world. To achieve this, its products must be an important commodity to trade.

ARAB MINERAL TRADE AND CO-OPERATION

At the beginning of this section it should be emphasized that the Arab mineral trade picture is much like that of the rest of the world. On the one hand it shows dominance by a few countries in possession or consumption of mineral resources. On the other hand, minerals are traded by agents and companies, and face the fluctuations that typify the international market. The Arab mineral economy continues to grow, but at a slow pace, and generally with low revenues.

No doubt there are many reasons for this, some to do with the world as a whole, others related to Arab factors. Worldwide, the mineral industry of the early 1980s witnessed a period of negative trends, with declining levels of production, trade, consumption and revenues (US Bureau of Mines, 1983). Several international political crises affected mineral output and flow, such as the Iran–Iraq war, Poland and Israeli–Arab antagonism. Many of these trouble areas are in the Arab World, from its western Moroccan boundaries (there are often conflicts in the Arab West), through Egypt–Libya and Sudan–Ethiopia–Somalia, the South Yemen upheaval, the Lebanese crisis and its link to the Israeli conflict, all the way to the eastern Iraqi boundaries, climaxing with the debilitating Iran–Iraq war. Purely Arab factors are the volume of their mineral production and its status in the world, the role mineral resources play in Arab industry and the Arab trade framework (AMC, 1981).

The Arab Mining Company report to the Fourth Arab Mineral Conference, held in Amman in 1981, stressed the extreme difficulty of acquiring useful inter-Arab statistical data on the mineral trade. This is a part of one of the concerns repeated throughout this book: the lack of proper Arab information exchange. Arab mineral production has been detailed earlier, and its impact on world markets shown to be tiny, except for phosphates and, to a much lesser extent, mercury and sulphur. It is not expected that this will change in the near future. The Arab mineral industry is not strong and inter-Arab transactions are very restricted. It is obvious that Arab mineral trade will become globally significant as long as there is no inter-Arab strategic policy. This policy should be mineral-

and industry-specific, and should distribute the centres of production and consumption logically throughout the Arab World. This would increase the volume of Arab mineral trade. The Arab trade market is at present dominated by short-term contracts. This gives higher revenues but instability; instead there should be long-term contracts, which give lower revenues but a more stable and continuous trade market. This cannot be achieved while there is a unilateral approach, competitive atmosphere and lack of information exchange.

The present times are difficult for the mineral resource sector, and the importance of mineral exports to the developing countries has diminished. The significance of exports of manufactured goods is rapidly increasing. It has become apparent that an adequate resource base exploited unilaterally does not ensure adequate domestic mineral production in the face of increasing foreign competition and diminished markets. This analysis was made by Van Rensburg (1986), who added that the differences among Third World countries (Arabs being typical) 'has proven a major hurdle to effective collective action bargaining with the industrial consumers'.

The Arab export–import figures in Tables 4, 5 and 7 and Figure 2 reflect the overall inefficiency of Arab mineral trade in the 1970s and 1980s. Too much competition, overlap, lack of co-ordination, foreign orientation, and short-sighted planning characterize the four regions. There will definitely be a poor future if Arab authorities do not try to complement each other, and current trends do not suggest that they will soon.

Figures 6 and 7 show the external trade and trading partners of Arab countries. Only three—Morocco, Jordan and Lebanon—have 30–35 per cent of their total imports made up of materials related to minerals, most of the rest import less than 10 per cent, and Libya, Egypt, Sudan, Djibouti, Oman and Qatar import less than 5 per cent. Only two countries have more than 50 per cent of their total exports made up of mineral-related materials (Mauritania and Morocco) and most of the rest of the Arab countries export less than 5 per cent. As long as Arab mineral development projects are carried on single-handedly, this picture will continue. Suppose a country X develops, for example, its copper deposits, and its next door neighbour, country Y, also develops its copper ores. X and Y will be handicapped by the costs of developing this mineral commodity. Each country will have to undertake full-scale exploration, prospecting, mining, processing and marketing. Would it not be logical to assume that Arab imports and exports would improve to a great extent if their efforts were complementary? Of course it would, but this requires the political will.

Figure 7 shows that most Arab countries secure their imports from foreign sources, their reliance on Arab sources being less than 15 per cent. Only Jordan, Lebanon and Somalia send more than 70 per cent of their total exports to their Arab partners, followed by North Yemen and the

UAE at 50–70 per cent, then Mauritania, Morocco, Egypt and South Yemen at 20–50 per cent, with the rest at less than 15 per cent. This crippled market is a logical consequence of the unilateral Arab policies followed. Short-sighted unilateral mineral development by definition leads to heavy reliance on foreign sources because the Arab market, which would otherwise have developed, did not have the chance to do so.

The supply–demand equation is the Arab use of and requirement for mineral matter. The demand for minerals is a derived demand, i.e. demand for finished products or for incorporation into a processing chain for other materials, and hence is substantially influenced by the level of economic activity (Humphreys, 1984). Two important points in analysis of current and future patterns of consumption are the criticality of supply and possible substitutions for minerals, whether partial or total (Desforges, 1984). Mineral processing technology is advancing at a high pace, and the huge indigenous mineral shortages in the advanced countries are forcing them to look for substitutes. This means that many of the Arab resources, both metallic and non-metallic, may not be of significance in the future. This applies to several elements that alloy with iron or aluminium, and to asbestos, magnesite, mica and certain clays, among others. In addition, the relative importance of primary products in global trade has fallen sharply since the 1950s, from 42 per cent in 1955 to about 20 per cent in the early 1980s, and the importance of manufactured goods has risen sharply (UNIDO, 1981c). However, the value of world exports of manufactured goods started declining in 1981, and the bulk was still concentrated in a small number of countries (UNIDO, 1983). About 80 per cent of the manufacturing exports of developing countries are from resource-based industries, and many of these exports have been transformed to a limited extent. The patterns of trade of developing countries showed an increase in imports of unprocessed materials from 14.2 per cent in 1970 to 23.9 per cent in 1980, while exports fell from 52.5 to 43.0 per cent. Imports of processed materials fell from 64.5 to 58.6 per cent, and exports rose from 24.4 to 38.9 per cent. This shows little improvement. The Arab countries are no better.

In no field does intra-company (and in the Arab case, intra-country) trade seem more prevalent than in the minerals business. The minerals sector can contribute to developing linkages between itself and the rest of the economy, and thus to an integrated economic structure. This integration requires public control over the resource base, and demands co-operation. A good approach is to establish a *functional* co-operative marketing organization, with powers to improve the economics of established processing facilities, to specify work inputs and outputs, to assemble experienced staff and to maintain the efficient implementation of co-operative projects. As nicely put by Bonar (1982), to co-operate is to

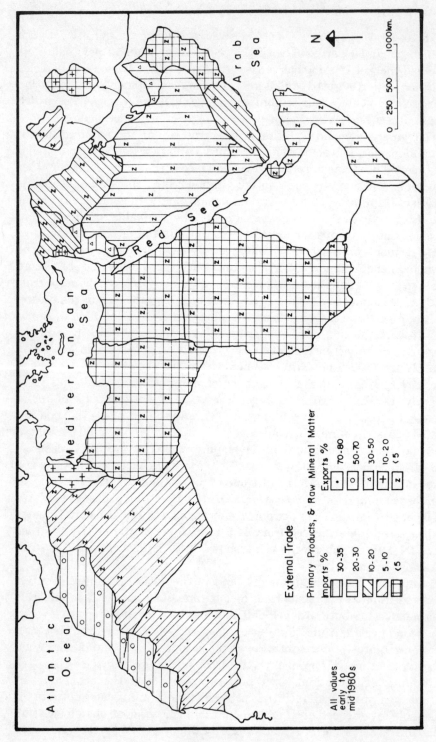

Figure 6. Contribution of primary products and raw materials to Arab external trade, by country.

External Trade

Primary Products, & Raw Mineral Matter

Exports %

●	70-80
○	50-70
◁	30-50
+	10-20
N	<5

Imports %

	30-35
	20-30
	10-20
	5-10
	<5

All values
early to
mid 1980s

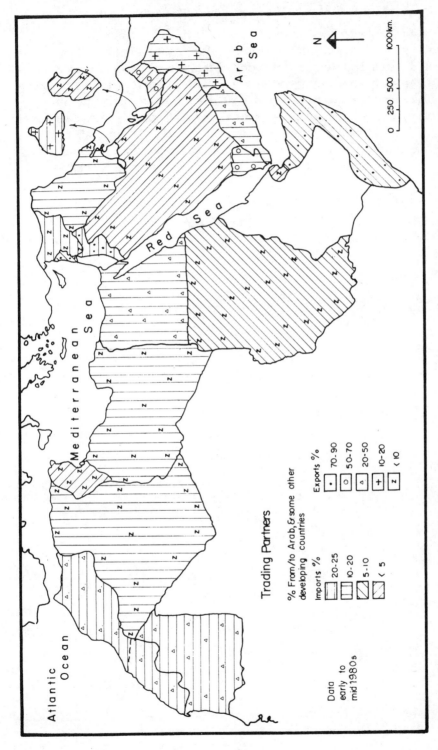

Figure 7. Proportion of Arab exports to and imports from other Arab countries and some other developing countries.

form a team, with each country a member linked not only by a common goal, but also by a common strategy. The question is how? The answer should concentrate on the following: Arab mineral co-operation requires realism, initiative and good management by partners, i.e. economic policies have to be viewed as a whole; there have to be structural improvements in public expenditure, tax systems, social security, etc.; everything related to inter-Arab trade, including movement of capital, products, labour and legislature, must be facilitated; trade relations among Arab countries and other developing countries must be strengthened.

III

MINERAL RESOURCES IN FUTURE ALTERNATIVES

He who foretells the future lies even when he tells the truth.

The future is always something to wonder about. How will this future affect Arab mineral resources and simultaneously, how will the resources influence the factors outlined in this book? From the start, and as implied in the above quotation, it should be understood that analysis of the future provides only possibilities, not something rigid, highly accurate and irrevocable. Given the turbulence of the world in general, and the broad spectrum of differences, events and situations in the Arab World in particular, future planning always leads to tenuous conclusions.

The importance of strategic planning for proper development cannot, however, be too strongly stressed. Plans have been based on forecasts, using production, exports, GNP and other such parameters, but in the present circumstances even the most sophisticated forecasts should be viewed with caution and even scepticism. An alternative to such forecasts is provided by scenarios. These are sets of credible, internally consistent but fundamentally different alternatives that portray future development environments (Galer, 1985). A scenario is simply an outline of a plot, and if it is reasonably acceptable it is a guideline for the course to be followed (Häfele, 1980). Scenarios are intended to challenge and alert decision-makers by providing concepts and descriptions of potential opportunities and threats.

It is only very recently, in the past 10 years or so, that specifically Arab studies tackling future developmental (or more commonly growth) aspects have emerged. Of course, many international studies covered the Arab regions. An excellent exposé of the shortcomings and lack of applicability to the Arab World of these studies was published by Sa'deddine *et al.* (1982). They described several global models, including those of the Club of Rome, Mizarovič and Pestel, Bariloochi Institute, Leontief, SARUM, UNIDO and the World Bank. The shortcomings are many, notably structural division of the Arab countries, or the grouping of some Third World countries with them, concentration on limited and rigid economic factors, and most importantly a failure to project the future of the Arab countries in a coherent framework. More significant was their exposé of several Arab studies dealing with the Arab future, contributed by individual scientists, government divisions, research institutes or organizations of the Arab League of Nations and its several summits on future Arab directives. The shortcomings of these Arab studies included: taking the present position (with all its negativities in terms of Arab co-operation) as a basis for future growth; a typical hesitancy over Arab co-operation, which cannot be changed simply by wishful thinking; uncertainty about the means of creating a better future; and a failure to consider local, regional *and* global development factors. Useful Arab contributions to this problem include those of: Sayegh (1985) on the determinants of Arab economic development; Nassar (1982) on Arab potential for alternative development; Al-Khatib (1984*a, b*) on projections of Arab future development in alternative frameworks; Al-Wahda (1985), Zurayk (1982), Eid (1977) and a number of special issues of Arab magazines and journals on various social, political and industrial aspects. There are several others. Some studies have concentrated on certain regions, such as those on the Gulf Co-operation Council, to which several referrals have been made previously in this book, or on the Arab West (Al-Manjara, 1983); and many on individual countries, such as Kuwait (Girgis, 1984) or Syria (Butterfield and Kubursi, 1981). There is an enormous amount of literature in the Arab World, and I cannot claim to have covered it all here.

Despite the volume of literature, no study has yet focused on the future role of Arab mineral resources. Even the various studies of the AOMR, which are full of steps and procedures to exploit the resources, do not really follow an approach directly linked to the future status of the Arab countries in various scenarios. They deal with the resources essentially as a separate entity, and hence do not consider their input–output relationship with future Arab alternatives. It should be emphasized, however, that the AOMR studies and those of the Economic Commission for Western Asia (ECWA) have contributed significantly to a broader understanding of Arab mineral resources development.

Most of the remainder of this chapter will examine three plausible scenarios of future Arab development in the period 2000–2020. The scenarios are constructed from: a logical analysis of what has happened in the Arab World; other studies related to Arab future growth and development; what is, in this author's opinion, the issue of ultimate importance, which is the welfare of the Arab World as a complementary block of coherent regions and how best mineral resources can serve the attainment of this goal. Creation of these scenarios requires the collection and interpretation of the many indicators obtained by international and local investigators. This study does not claim to follow a unique path, but in this case a subjective analysis is felt to be necessary and scientifically valid. By their authors' own admission, the different quantitative scenarios studies have many shortcomings and are extremely complex (World Bank, 1984; UNIDO, 1981a, b, c; UNITAD, 1981; Butterfield and Kubursi, 1981; Marzouk, 1984; Nassar, 1982; Sa'deddine et al., 1982).

Since development of the Arab World is linked, in one way or another, to other countries, and specifically to the developed countries, high and low growth cases will be considered. The former assumes that the developed countries undergo a recovery and steady expansion, with the developing countries also benefiting from this recovery. The low growth case assumes that the economies of the developed contries will not improve, and hence the developing countries will feel the pinch. Next, the effect of the future scene on the performance of the mineral resource sector, either positive or negative, is shown within the changing balance of the scenarios and their two cases. How do the 'building blocks' (the influencing factors in this study: human, policies, administrative and financial) stand together, well or badly?

ALTERNATIVE SCENARIOS

The variables upon which the scenarios are constructed, as indicated above, are the trends showing up in the policies of the Arab countries and in the relections of various Arab development analyses. The scenarios are thus alternative possibilities, each of which could lead Arab development, and hence the mineral sector, along a completely different path. From these collective trends three alternative scenarios emerge: the Trend Scenario, the Improved Scenario and the Transformed Scenario.

The Trend Scenario

The Trend Scenario is the present ongoing situation. The plot assumes that the emphasis in Arab development will follow limited, mostly authoritarian, individualistic, single country or single sector growth, or at most some forms of close neighbour (intra-regional) co-operation not necessarily resulting from deep intent and long-range planning. Inter-Arab (Arab League) organizations play essentially a symbolic role, though with some technical consultation. The political scene is quite diverse, with clashes between countries.

The Improved Scenario

The plot under the Improved Scenario shows partial Arab development through greater, more intricate, co-operation on a bilateral or multilateral basis within or across regions, resulting from encouraging previous experiences. There is only immediate co-operation and mutual interaction in various related economic aspects, and several economic options are available, with a freer private sector and limited sharing of benefits. Inter-Arab organizations acquire an active but indirect role, still mostly of a technical nature. The political scene is more uniform but not free from polarization.

The Transformed Scenario

In the Transformed Scenario total Arab development is the goal, and is achieved through coherent Arab blocks (the four regions) linked by a well directed and active organization (Arab League) into complementarity on socio-economic and political bases. There are strongly planned ties between countries of the same region, especially socio-economically, and there is a strongly planned mutual relationship between the regions based on socio-economic and political aspects. Governments that are representative of the population are in control, and the private sector shares in contributing to the welfare of the country and the whole region. The political scene is one of coherence and complementarity, signifying Arab solidarity and stability.

REFLECTIVE INDICATORS

It is clear that the economies of the world are becoming more inter-dependent and that the role of developing countries is increasing in importance, but the development of international co-operation has not kept pace with these changes (Paye, 1985). One indicator that can be taken as an example is the 'energy coefficient', which is the percentage increase in energy use required for a 1 per cent increase in GDP. The developed countries show a value of 0.8 while the developing countries show a value of 1.5 for the year 2020 (Häfele, 1980). The developing countries still need to build up their infrastructures, considerable inequities still remain between the two groups, and the gap will continue to be a problem far into the next century.

In May 1985, the new chairman of Rio Tinto Zinc (one of the mineral giants) observed that 'the mining industry's longer-term outlook is much brighter. In the early 1990s most metal supplies will be broadly back in balance, and demand will continue growing. Yet investments are virtually dried up, therefore because of the long lead times needed for development of major new mines, decisions must be made within the next year or so for the next generation of successful mining projects' (*Financial Times*, 1986). Yet the Lima scenario and the Global 2000 Report give very similar years for the exhaustion of reserves of aluminium and iron (in about 2035) and of copper, lead, nickel, tin and zinc (in about 2005) at present consumption rates (UNIDO, 1981*a*, *b*). This consumption is currently, and is expected to remain, that of the developed industrialized world. However, the origin of these, and several other, mineral resources is increasingly the developing countries. The future link is obvious, and any recovery (high case) or depression (low case) will affect both groups. The World Bank (1984) considered the effects of the high and low cases on the annual growth rates of the GDPs of developing countries, and produced figures of 5.5 and 4.7 per cent, respectively. As a single indicator cannot supply a good interpretation of the future, many indicators for the Arab World have been obtained from various sources (shown in Table 15). Unfortunately, these indicators are not available for every Arab country, so they cannot be combined on a regional basis. All the sources used here group the Arab countries according to income level, oil export and import or geography. However, the excellent research done by Sayegh (1985) gives a deep insight into the future possibilities for the Arab countries, or at least the major ones. Sayegh's work is thus used to map out the regional scenarios.

Now is the place to make certain observations concerning these indicators (Table 15). If the high and low cases are compared, they do not show a considerable difference. The developing countries are apparently squeezed between a high growth rate for the North, which generates

Table 15
DEVELOPMENT INDICATORS FOR THE ARAB FUTURE (1985–2000)

From World Bank (1984)

1. Average annual percentage change in export *prices* for minerals and metals: 1.5

2. Average annual percentage change in export *volume* for metals and minerals: 0.5

3. Average annual percentage change in performance:

	High case	Low case
GDP	5.5	4.7
Export growth	3.4	2.5
Import growth	7.2	5.1
Merchandise export	5.2	3.7
Merchandise import	5.7	3.6
Export manufactured goods	9.4	7.2
Export primary goods	3.2	2.2

4. Current account balance and financing (in $billion at 1980 prices):

	High case	Low case
Interest on medium- and long-term debt	− 7.4	− 8.1
Current account balance	− 17.3	− 12.8

5. Growth of labour force: 3.4%
   Population projection: 292 million

Table 15—*continued*

From UNIDO (1981b)—Lima target

1. Annual growth rates and selected sectoral value added:

	GDP	Industry	Services
	6.56%	5.58%	8.66%

2. Labour demand and supply by sector:

	Industry	Services
Demand	15.6 million	27.7 million
	(23.3%)	(41.4%)
Supply	17.5 million	31.2 million

3. Share of world MVA:

Reference Scenario (R)	3.0%
Third UN Decade Scenario (3UN)	3.7%–3.9%

4. Percentage average annual growth rate of value added by sector:

	GDP	Mining	Manufacture	Others
R	7.4	4.3	11.0	8.8
3UN	6.6	3.8	9.4	8.4

5. Percentage of GDP total value added by sector:

	Mining	Manufacture	Others
R	46.1	7.9	39.2
3UN	22.6	19.0	54.6

6. Percentage share of world GDP and exports:

	GDP	Exports
R	5.0	10.0
3UN	4.8	7.1

From UNIDO (1981b)—The Trend Scenario

1. Protectionism in developing countries: same
2. Technology: still adopted from developed countries
3. Growth rate: GNP 5.9%, *per capita* 2.9%
4. Total unemployment: 17 million
5. Accumulated investment rate at 1981 prices: 21.6%
6. Basic balance: $174 billion
7. Outstanding debts: $199 billion

From Sayegh (1985)—Collective total analysis

1. Growth potential:

Arab West	poor to excellent
Nile Valley	poor to fair
Fertile Crescent	fair to very good
Arab Peninsula	fair to good

2. Requirements:
All need to improve democracy, welfare and technology

unbearable tensions, and a low growth alternative in which their earnings are shrinking. This calls for a reorientation of trade and finance among the developing countries (UNITAD, 1981). The export indicators in Table 15 are not all that encouraging. The financial picture does not look too bright, as in both the low and high cases there are negative values for interest on debts and the current account balance; there are also high accumulated investment rates and huge outstanding debts. Moreover, the Lima target is out of line, in that 25 per cent of the world's industrial production will not be secured by the developing countries by the year 2000. Indicators in Table 15 for industry and manufacture in the Arab countries are quite variable in the different models, but overall they are better than other economic aspects, and the same applies to the mining and services sectors. However, there is high unemployment, which is further worsened by a low growth in demand for labour and a population explosion. The protectionist policy that would still be followed by the developing countries and the continuing dependence on adopted technology shown in Table 15 would not lead to very encouraging outcomes.

Nowadays, many journals, magazines, surveys and conferences are pointing to a revival and expansion of development and growth in the developed countries, though at a slower rate than before. It is believed, and reflected by the above indicators, that the effects of this recovery on the developing countries are not likely to be very important (positively) in terms of prices, manufacture, trade and finance (Urquidi, 1983). The industrial countries will remain inward looking and hence the developing countries need more rational policies to promote self-reliance and to upgrade their economies.

It is always the welfare of Arab society within a policy of complementarity that should be aimed for. The analysis given by Sayegh (1985) shows the Arab regions to vary from poor to excellent in their potential for growth. They all have very demanding requirements, and so must re-orient their trade and finance, and follow more rational policies that *promote self-reliance to achieve the desired welfare.*

This analysis must not ignore a very significant aspect now emerging as a result of rapid advances in materials technology and the recycling of metals. This will place another hurdle before the economies of developing countries. There has been a quantum leap in the development of metal composites, polymers, glass fibres and ceramics. Third World policy-makers are finding that their mineral exports are being threatened by less material-intensive manufacturing processes (Steele, 1988). In 1981–82, 84.2 per cent of the lead, 37.9 per cent of the copper, 27.2 per cent of the aluminium, 24.2 per cent of the zinc and 19.0 per cent of the tin used in the West was recycled. Many experts in the mineral industry sector are suggesting that the 'dematerialization' of manufacturing processes is reaching the point where mineral commodity markets are unlikely to return to their previous high price levels even if world economic growth picks up. The Arabs are therefore forced to put their mineral resources to good use now, before it is too late. They need to develop these resources.

REQUIREMENTS FOR MINERAL RESOURCE DEVELOPMENT

The three scenarios are the possible plots, and the indicators are the guides. So performance in mineral resources is best observed by looking at how the factors influencing mineral development—human, policy, administrative and financial—will do in the different scenarios. This needs a list of the requirements for a fully functional Arab mineral development as they have emerged from analysis of the factors detailed previously.

General Requirements

Regardless of which influencing factor one considers, the seven foremost requirements needed are:

- Arab information exchange in a systematic standardized manner.
- A binding political decision to link the development process within an Arab continuum.
- Permanent, binding co-operative action to implement decisions.
- An Arab legislative code of practice that will harmonize and control these co-operative actions within the framework of total Arab welfare.
- Giving the AOMR and its subsidiaries enforcive powers in regional mineral projects with unswerving priority.
- An Arab technological revolution.
- An Arab cultural revolution, preserving the traditions that tie the Arab countries together, but improving those social practices that are holding Arab society back, notably with regard to democracy.

Specific Factor Requirements

Human

Control the population explosion, and redistribute the working force to exploit mineral resources in remote underdeveloped areas.

Reduce reliance on non-Arab manpower in the mining industry, with technical training and institutional upgrading.

Re-orient some higher education institutes to focus on mineral processing.

Restructure trading towards import of raw materials and export of finished products as this induces a change in the quality of the working force.

92

Policies

Adopt strategic policies on mineral development that engender collective Arab welfare regardless of changing governments.

Establish vertical and horizontal sharing in the decision-making process in government geological divisions and other mining enterprises.

Create corporate mining organizations and encourage small-scale mining, especially in remote areas.

Strengthen non-financial functional ties between Arab countries in mineral involvements.

Give priority to Arab mineral enterprises.

Ensure that international mining enterprises' involvements with Arab countries are more mutually beneficial.

Strengthen mineral involvements with other developing countries, and with advanced countries that do not have a manipulative approach, and emphasize interaction with the UNDP or its equivalent.

Administrative

Upgrade and standardize existing Arab mining codes.

Link the mineral-related government decision-making authorities to those of the mineral industry and research, and place highly qualified personnel in key positions.

Upgrade, standardize and merge mining institutes in neighbouring countries.

Continuously assess mineral resources, with wider and better exploration.

Financial

Homogenize Arab financial actions towards minerals development.

Create an efficient absorption of Arab finance in the mining sector.

Define mineral development priorities and put the Arab minerals' potential into actual economic development.

Minimize mining authorities' bureaucratic interference.

Strengthen and functionalize the AOMR Revolving Fund.

Adopt a co-ordinated mineral industry plan that is on the right scale, and thus economically feasible, at the regional level and then at the total Arab World level.

Spread mineral industrial involvement to all Arab countries, especially promoting local small-scale industries in remote areas.

Strengthen integrated mineral industries within a country, a region and the total Arab World.

Create mining industrial centres that serve more efficiently the spread of mineral resources among the Arab countries, thus eliminating competition.

Enact codes to regulate the exchange of capital, products and labour in the mining sector of the Arab countries.

Adopt an Arab mineral exchange policy to strengthen specific minerals and industries, with a fair geographical distribution.

Enhance the creation of an Arab market for minerals.

Replace short-term contracts with long-term contracts, inducing a more stable mineral market.

Eliminate overlap and competition between the Arab countries and create a collaborative chain-type market.

SCENARIO PERFORMANCES

Given the general and specific requirements, it remains to be seen how they would fare in the three scenarios. First, it should be stated that since the indicators in Table 15 show little difference between the high and low cases, it is the inter-Arab (by country or region) relationship that is going to be the decisive factor in the overall future development. It was shown that in both cases the developing countries, including the Arabs, would suffer a worsening situation compared to the developed countries. Hence, the more the Arab countries can reduce their dependence on foreign sources, and the more they complement each other, the better is their chance of a developed future.

The Trend Scenario

The future for this plot looks pretty much the same as the present, with a possibility of deterioration. None of the general requirements is met. The human situation is worse, especially because of the population explosion and associated unsatisfactory socio-economic conditions. None of the policies requirements is achieved, although there may be glimpses of corporate organizations here and there. Only partial mineral assessment appears. Finally the existing non-homogeneity in Arab actions could worsen, and this goes with a non-absorptive, non-integrated, shattered Arab market dominated by short-term contracts and financial aid for the mineral industry or mineral trade.

The result is *restricted mineral growth* for a certain time interval. In view of the international competitive market, it would not last long, to the year 2000 at the latest.

The Improved Scenario

This holds better promise with partial real development seeping through. At the general requirements level, information exchange takes place bilaterally and, to a limited extent, on a collective scale through the AOMR, with some technological mutual benefit from the implementation of certain mineral resource projects. No restructuring or basic reorientation of Arab technical manpower is seen except for short intervals of time and for specified mining projects, helped by the available training institutes which reduce, partially, reliance on non-Arab manpower. In terms of the policies requirements, corporate organizations appear, through private sector encouragement, but face some difficulties as they follow short-range planning with a limited populace sharing in the decision-making process. There are links between a few government mineral authorities and the mineral industry, enhanced by research directives. Groups of two or more countries work out mutual priorities in mineral resource projects, especially with the help of the AOMR and AOID. This creates a better technological assessment of resources and the upgrading of mining codes. The financial scene is partly homogenized and dictated by the limited mutual benefits expected from a small number of co-operative projects. There is some absorptive capacity, which leads to an enhanced mining sector, and increases its contribution to social welfare. The AOMR Revolving Fund partially replaces, at least on a regional scale, some of the other Arab funds, thus strengthening the AOMR's position. This applies more to projects of large scale to justify their economic feasibility. Some attempts to build mutual codes for exchange of selective Arab commodities strengthen the chances for micro-Arab markets, especially through 'islands' of integrated chain industries among two or more neighbouring countries. These economically created neighbour blocks tend to form polarized groups that try to secure the best opportunities for themselves, regardless of possible negative effects on other Arab countries.

The result is *widespread mineral growth* for a certain time interval. If the international mineral market jumped positively forward and retained gains made, this approach would tend to converge with a developed framework. However, the international mineral market is not expected to jump, and it is more likely that growth would come but at a slower pace. Hence, Arab mineral growth would be hindered by the international market, and would be very likely to dwindle to unimportance by 2010 or so.

The Transformed Scenario (or the Desired Future)

This is the desired plot in which the welfare and real Arab development goals could be attained. In fact, this could be the scenario which the Arab

World is heading for *after* the Arab countries have 'exhausted' the other two scenarios and found, in a manner exemplary to other regions in the world, that complementarity is the only path to self-reliance and promotion.

In this scenario a culturally, technologically and legislatively Arab-oriented uniform and homogenized mineral development plan prevails through full information exchange procedures, regularized and ordained by the AOMR, and requested by Arab governments who have taken a binding decision to implement. The general requirements are met, though in a gradual manner.

On the human front, the steps taken in restructuring Arab commodities and in education and technical training lead to the replacement of a great percentage of non-Arab manpower. With new cultural concepts spreading and the welfare of society increasing, population growth is controlled and, most significantly, many people inhabit the inner remote areas and exploit and develop their mineral resources. This is possible only because the mineral resource policies followed at country level are part of a coherent plan at regional level that has defined the priorities of that region taking account of socio-economic benefits to the member countries and their intermingling populations. Large trans-Arab mineral companies are dependent on the contribution of widespread corporate mineral organizations with small-scale mining operations largely encouraged by the mining authorities. These authorities have upgraded and merged many intra-regional geological services as a result of common research and extensive mineral assessment, aided by uniform mining codes. The AOMR and its subsidiaries ensure that inter-regional mining projects are complementary in terms of exchange of technical personnel, experience, labour and mutual economic benefit.

The greatest achievement in this scenario is in laying down the basic structure for a stable Arab market, which grows slowly but steadily thus eliminating competition and strengthening homogeneity. Although the positive impact of this market takes some time to be seen, because of the binding decisions of the Arab authorities and the co-operating efforts of the Arab League organizations it will catch up at a fast rate. This stable market has the capacity to absorb centrally integrated large-scale chain industries supplemented by an even geographical distribution of small-scale industries. Inevitably this actualizes the Arab mineral potential into an economic blessing that is oriented towards the welfare of the Arab people. This should not be too far-fetched as the AOMR Revolving Fund has the financial capacity to give long-term aid, and can also intercede throughout the Arab World to provide the needed expertise as well as the market.

The result is a *developed mineral sector* that has been created mainly by intra- and inter-Arab co-operative practices within coherent regions and a complementary Arab World. International fluctuations or irregularities in prices, volumes, availability and other areas of the mineral scene could be absorbed by the total Arab block. Most importantly, the Arab contribution to the world mineral scene would be positive, stabilizing and constructive. Such a scenario would not only enhance the development of an Arab country, Arab region or the Arab World, but would extend into development of the whole world.

If this plot *is* the inevitable path leading from the other two scenarios, it cannot be expected before the second or third decade of the twenty-first century, around 2020. It is unfortunately true that many of us will not live long enough to witness this transformation, but coming Arab generations could enjoy a developed society whose mineral resources have been a blessing. Is there a short-cut that the Arab authorities can take to this desired future?

CHALLENGES OF THE COMING DECADES

'Those people do not own oil, they are only sitting on it.' This was said by the Secretary of the Treasury of the USA in 1976 about the Arabs. After what has become known as the 'oil crisis', certain mechanisms were followed by those dominating the international market that prevented the Arabs from fully using their oil wealth. In the 1970s and early 1980s the Arabs bought their present, but unfortunately at the expense of their future. Where do oil revenues stand now in view of the deteriorating status of the Arab World: decreasing economic growth, increasing debts, increasing imbalance between productivity and consumption, and other negative indicators? The question that should now be asked concerns the future of Arab mineral resources as an alternative to oil: Will the Arabs just 'sit on' these resources, or will they exploit them properly?

The underlying theme of this book is obvious, and that is that these resources must be properly utilized in good time within a co-operative Arab plan among the Arab regions and countries within these regions, under the auspices of the Arab League (notably the AOMR). The resources should be used to serve the well-being of Arab society within a framework of real Arab development. The driving force is also obvious: co-operation in developing the mineral resources within a unified well defined Arab plan, taking the requirements detailed in this book into consideration. These requirements demand that the following proposals be implemented:

1. Control the population growth; build up the Arab workforce in the minerals domain; improve and re-orient some of the existing insti-

tutes, population concentrations and trading commodities to induce a qualitative change in the mineral-related workforce.

2. Adopt and apply a strategy for proper exploitation of Arab mineral resources through information exchange, technology and interaction among regions' and countries' specialized Arab institutes, with the aim of benefiting all of society.

3. Enact and apply mineral resource codes and regulations to homogenize and standardize Arab mining activities spread evenly among neighbouring Arab regions and countries, facilitated by Arab manpower and technical exchange, so that benefits can be shared equally.

4. Strengthen and solidify the flow of long-term Arab finance to mineral resources; upgrade the industrial transformation of these resources (large-scale at regional level and small-scale at country level) to serve an inter-regional priority Arab market, capable of absorbing the Arab capital and commodities and necessarily oriented towards basic development of the whole social framework, and not towards a consuming elite.

5. Establish and strengthen long-term mutual relationships among Arab countries of the same region and among regions, eventually leading to complementarity and coherence under the Arab League. This should open the way for better and more equal relationships with foreign (non-Arab) countries or regions.

If the Arab countries start to implement the above proposals and the details related to the factors (human, policies, management, finance) influencing their mineral resources and their development, then it is quite possible for the Arab World to make a positive leap towards helping its own people and other peoples of the world. The scene depicted under the Transformed Scenario could even be realized before 2020.

This study has stressed that complementarity and coherence among the four Arab regions (each enjoying its unity) are the key elements in the development of mineral resources, and therefore in the development of the Arab World. Thus the natural question is: how far have the Arabs gone towards this complementarity?

During 1989 all the peoples of the world looked towards Western Europe and wondered what it would be like in 1992, when a 'unified Europe' is promised. At the same time, the world superpowers, the USA and the USSR, are coming closer to each other. The Europeans started with a common market and are heading towards complementarity and coherence. Are the Arabs learning?

The most recent (early 1989) 'blocks' of Arab countries coming together in various forms of unity are what has been referred to in this study as the

Arab West (Region A: Algeria, Libya, Mauritania, Morocco and Tunisia), which is called the Federation of the Arab Maghreb (*maghreb* meaning west), and the Arab Co-operation Council, comprising Iraq, Jordan (from Region C), North Yemen (from Region D) and Egypt (from Region B). So apparently the Arabs are learning. However, this is not the first time that Arab countries have come together in somewhat poorly defined groupings, which break up after a short time. There are several examples. Egypt and Syria in 1958 is the earliest, but a number of unions of very short duration followed (some simply being announced without actually being realized) between Syria and Iraq, Egypt and Libya, Libya and Tunisia, Egypt, Syria and Libya, Egypt and Sudan, the two Yemens, etc. The one that has persisted, although it was established in 1981 as a counter to the Iraq–Iran war, is the Gulf Co-operation Council (GCC) between Saudi Arabia and the small Arab states stretching along the Arabian–Persian Gulf (Kuwait, Bahrain, Qatar, UAE and Oman, all of Region D); this organization is suspicious of both Iran and Iraq. All these 'unified' entities, and especially the two most recent ones, were a response to some external pressure. They did not arise from a well intended and projected plan towards complementarity and coherence, although in the case of the Arab West it might work out that way in the future.

This study is not intended to be politically oriented, and therefore any comment regarding blocks of Arab countries approaching each other in some form of unity is only related to how much they comply with the thesis of the study. Co-operation (for the development of mineral and other resources) is the name of the game; but this co-operation must be authentic and purposefully intended as a priority for the benefit of all the people within the same region. Common interests among different countries of different regions are definitely real, but they are not as authentic and purposefully intended for the welfare of people as the common interests of sharing a neighbourhood that has shaped their very existence. It is after a region has built itself up through the co-operation of its member countries that it can and should extend itself to other regions. This follows a deep and well analysed plan for unity within a region, and complementarity and coherence among the regions.

The two most recent examples of Arab countries coming together to form unified blocks have their connotations within the context of the thesis of this book. We can divide the Arab World along new lines as follows:

1. Those countries essentially still in the Trend Scenario are Somalia, Djibouti, Sudan, Syria, Lebanon and South Yemen.

2. Those essentially at the margin between the Trend and the Improved

Scenarios are the countries in the newly formed Arab Co-operation Council.

3. Those essentially at an early stage of the Improved Scenario are the countries of the GCC, provided that they solidify their unity, democratize their regimes and diversify their economies.

4. Those apparently at an early stage of the Transformed Scenario are the countries of the newly formed Federation of the Arab Maghreb, also provided that they further solidify their unity, democratize their regimes and extend benefits to other Arab entities.

Finally, returning to our concern with Arab mineral resources, and assuming that the Arab countries fit into the above categories, it should be apparent that these resources will fare in different ways. The better chance for Arab mineral resources to make a decent contribution to real development of the Arab World occurs where there is a framework of countries taking steps towards the Transformed Scenario. If the Arab West is actually following this path then it would be expected that its mineral resources will be exploited far more fully than they are now. The other countries, some of which have important existing or potential mineral resources (e.g. Sudan, Egypt, Somalia, Yemen, Saudi Arabia, Jordan, Syria and Oman) should strongly seek out their regional counterparts to build up proper unity, or their mineral wealth will just be another illusion.

APPENDIX

Geological Synopsis of the Arab World

No single systematic set of geological maps covering the whole area existed before 1987, although many meetings on this were held and recommendations were made a long time ago (Khawlie, 1983*a*). One obvious reason for this situation is poor co-ordination, as the production of the map was being 'grabbed' by one organization or another, including the Arab Geological Association, the Arab Organization for Mineral Resources, the Arab League Educational Co-ordination Organization and the Organization of Arab Petroleum Exporting Countries. Difficulties arise from the great variability in quality, scale and coverage of existing maps, but especially from non-standardization of stratigraphy, lithology and terminology of units. This variability is caused by most, if not all, basic maps' being prepared by different foreign geologists (French, British, Italian, Russian, German and American) in the different Arab countries. The recent maps of the geology and mineral resources of the Arab World (Collinson *et al.*, 1987) should be warmly welcomed when released to researchers, professionals and the public.

There follows a briefing on the geology of the Arab countries in Asia, i.e. east and north of the Red Sea, and in Africa (Figure 8). The Asian geology is adapted entirely from Beydoun (1988), to whom the reader should refer for more details.

ASIA

The Asian side comprises the Levantine plate (Sinai and the maritime areas of Syria, Lebanon and Palestine) and the Arabian plate (the remainder), including south-east Turkey and south-west Iran, bounded by the Taurus–Zagros thrust zone. Basement Precambrian crystalline rocks and metasediments cover the Arabian shield extensively with limited young extrusives in the north and centre. The shield evolved in an island arc environment from extensive volcanism and cannibalistic sedimentary deposition accompanied and interrupted by tectonism, intense plutonism and metamorphism. This took place in three episodes, constituting the

101

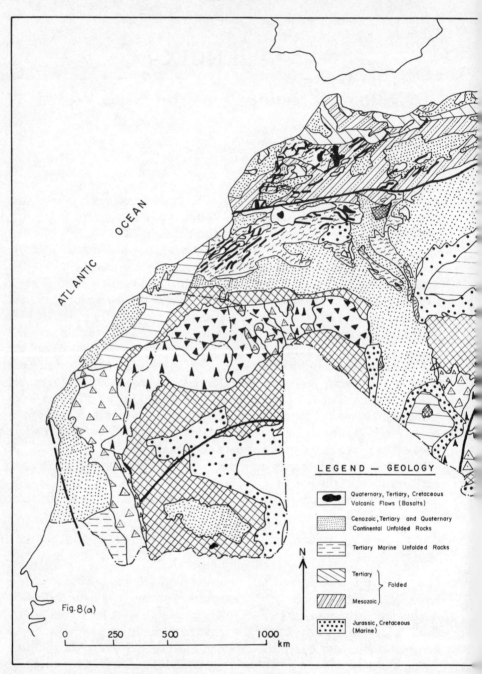

Figure 8. Simplified geological maps of the Arab World.
(a) Region A; (b) Region B; (c) Region C and D.

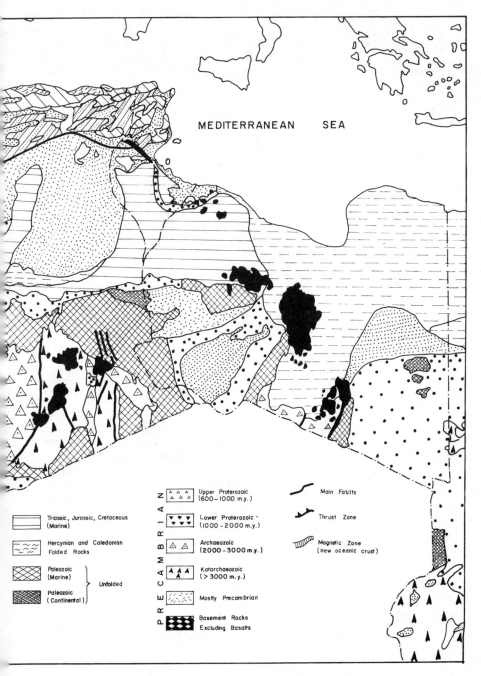

MEDITERRANEAN SEA

Triassic, Jurassic, Cretaceous
(Marine)

Hercynian and Caledonian
Folded Rocks

Paleozoic
(Marine) } Unfolded

Paleozoic
(Continental)

PRECAMBRIAN

Upper Proterozoic
(600–1000 m.y.)

Lower Proterozoic
(1000–2000 m.y.)

Archaeozoic
(2000–3000 m.y.)

Katarchaeozoic
(> 3000 m.y.)

Mostly Precambrian

Basement Rocks
Excluding Basalts

Main Faults

Thrust Zone

Magnetic Zone
(new oceanic crust)

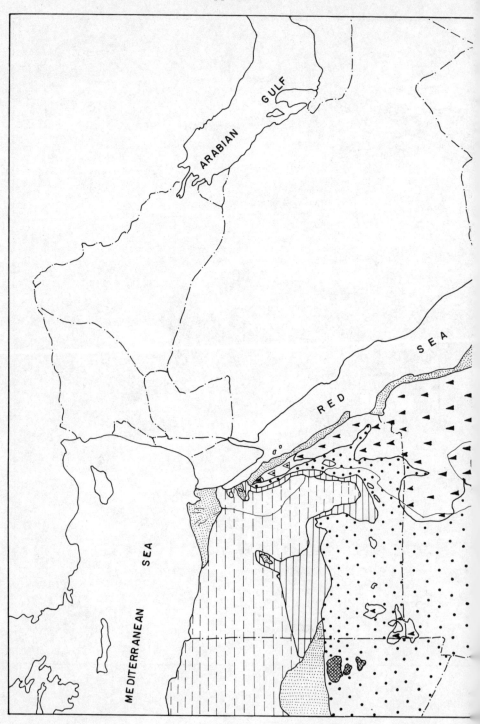

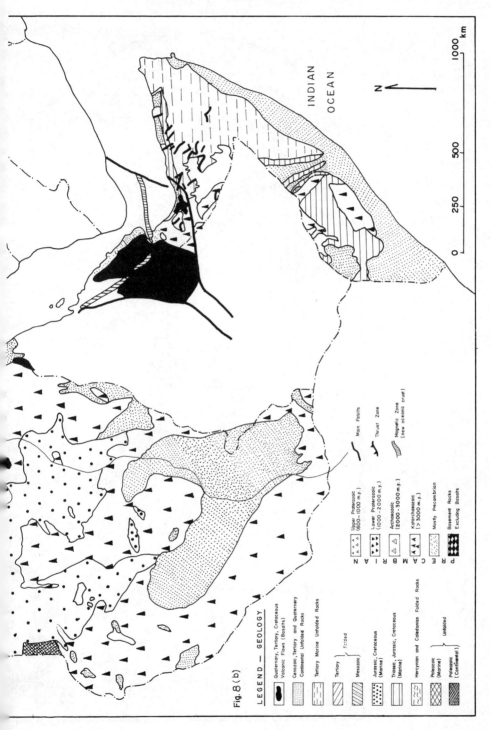

Fig. 8 (b)

LEGEND — GEOLOGY

■	Quaternary, Tertiary, Cretaceous Volcanic Flows (Basalts)
	Cenozoic, Tertiary and Quaternary Continental Unfolded Rocks
	Tertiary Marine Unfolded Rocks
	Tertiary } Folded
	Mesozoic }
	Jurassic, Cretaceous (Marine)
	Triassic, Jurassic, Cretaceous (Marine)
	Hercynian and Caledonian Folded Rocks
	Palaeozoic (Marine) } Unfolded
	Palaeozoic (Continental) }

Upper Proterozoic (600-1000 m.y.)

Lower Proterozoic (1000-2000 m.y.)

Archaeozoic (2000-3000 m.y.)

Katarchaeozoic (>3000 m.y.)

Mostly Precambrian

Basement Rocks Excluding Basalts

PRECAMBRIAN

Main Faults

Thrust Zone

Magnetic Zone (new oceanic crust)

INDIAN OCEAN

N

0   250   500   1000 km

105

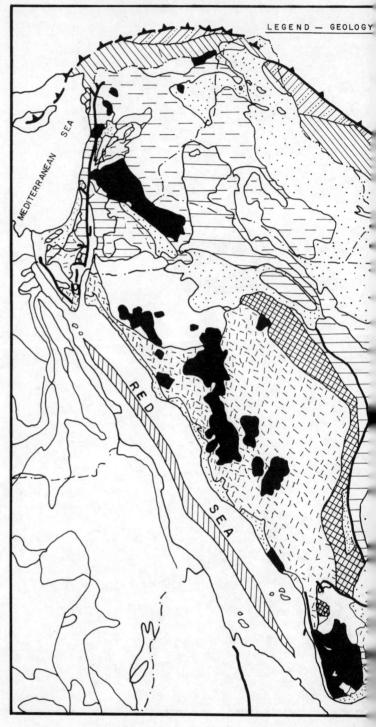

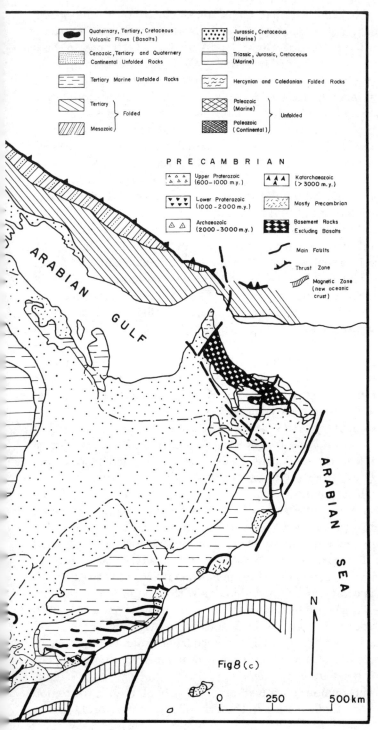

Quaternary, Tertiary, Cretaceous Volcanic Flows (Basalts)

Cenozoic, Tertiary and Quaternary Continental Unfolded Rocks

Tertiary Marine Unfolded Rocks

Tertiary } Folded
Mesozoic }

Jurassic, Cretaceous (Marine)

Triassic, Jurassic, Cretaceous (Marine)

Hercynian and Caledonian Folded Rocks

Paleozoic (Marine) } Unfolded
Paleozoic (Continental) }

P R E C A M B R I A N

Upper Proterozoic (600–1000 m.y.)

Lower Proterozoic (1000–2000 m.y.)

Archaeozoic (2000–3000 m.y.)

Katarchaeozoic (>3000 m.y.)

Mostly Precambrian

Basement Rocks Excluding Basalts

Main Faults

Thrust Zone

Magnetic Zone (new oceanic crust)

ARABIAN GULF

ARABIAN SEA

N

Fig 8 (c)

0   250   500 km

107

Hijaz tectonic cycle. The cratonization of the island arc went through an immature, mature and then continental crust with a switch in magmatism from calcalkaline to peralkaline. Eight depositional groups belong to three main phases, and correspond to the Hijaz cycle. Phase I, extending from 1200 to about 900 million years, reflects a moderate to shallow marine environment overlain and underlain by deep marine flysch environments. Phase II shows three distinct groups: volcanic (andesites and rhyolites) in the middle and molasse carbonate shelf sediments at the top and bottom, extending to about 640 million years. Phase III is predominantly terrigenous clastics but with a substantial Jubaylah carbonate group, extending in time into the Early Cambrian. Further south in Yemen, there seems to have been a similar evolution, and the basements in Somalia and Dhofar are associated with the Arab craton.

Infra-Cambrian thick non-metamorphosed alternating clastics and carbonates (Huqf group) overlie basement in south-east Oman and Dhofar, and correlate with exposures in northern Somalia, and with phase III. The Proterozoic–Phanerozoic boundary was marked by continuous sedimentation covering much of eastern and low-lying western Arabia. An interesting paleoclimatic indicator, Proterozoic tillites, are found in Oman. Arid conditions followed as the Huqf group terminates with evaporitic halites.

Syngenetic stratabound or hydrogenetic and epigenetic minor deposits are associated with the volcanic and plutonic sequences, or shear zones of the craton. Notable are mineralizations of copper, zinc, lead, gold and silver, fluorite, REE, bauxite, pyrite, nickel and cobalt.

Palaeozoic sedimentary deposits lapping the eastern edge of the shield extend from southern Jordan to Yemen in the south, with some discontinuous patches towards the Taurus–Zagros mountains, and in the Oman mountains. They are predominantly clastics with limited carbonates at the base, although carbonate deposition replaced the clastics towards the end of the Palaeozoic, represented by the widespread Upper Permian Khuff formation. In portions of the Oman mountains, Ordovician deposits show recumbent folding, overthrusting and low-grade regional metamorphism. These Palaeozoic features reflect positive land masses in the Shield supplying sediments to the epicontinental seas to the east with transgressions and regressions, terminating with a major transgression. Two glaciation records are evident in Palaeozoic Arabia: one Early with extensive evidence all the way to the Mauritanian–Algerian Sahara, and the other Permo-Carboniferous with evidence from Oman to Ethiopia.

No metallic or non-metallic mineral resources of any significance occur in the Palaeozoic deposits of the region.

Predominant carbonates continued into the Early Mesozoic, the region being low-lying with shallow seas. In the Triassic, there were arid to semiarid conditions with gypsiferous, coloured shales and sandstone continental deposits, and periodic limestones and interbedded dolomites-anhydrites. In the Late Triassic the south Tethys Ocean opened to the east and north, separating Iran and Turkey following uplift and rifting. The Levantine region formed the western continental margin of the Arabian plate, and linked to the north-spreading centre by a transform fault paralleling the Levant coast. The Mesozoic thus witnessed extensive carbonate banks and coral build-ups' notably with Jurassic transgressions over shallow waters from the Levant to Arabia and Ethiopia–Somalia. Towards the end of the Jurassic there were epeirogenetic oscillations, uplift, erosion, clastic sedimentation and interbedded evaporites (Gotnia or Hith anhydrite formation), believed to be laid down in giant sabkhas. The terminal Jurassic was a period of differential vertical uplift of blocks over the Levant and Arabia, with local volcanicity heralding the continental deposits of the Cretaceous. In this period the Nubian continental sandstone was deposited into the surrounding subsiding basins around the peripheries of the Shield, to change by the Mid to Upper Cretaceous into pelagic open sea carbonates of argillaceous limestones, marls and chalks.

Significant tectonics characterize the end of the Mesozoic with ophiolites and radiolarites emplaced from Cyprus and north-western Syria through the Taurus–Zagros into Oman. This marks a phase of subduction, collision and compression. The terminal Cretaceous saw tensional fracturing with flood basalt outpourings continuing intermittently in the Palaeogene and culminating in the rifting of the Gulf of Aden and the Red Sea in the Oligocene. Sea floor spreading began in the Miocene.

Other than the outstanding Mesozoic hydrocarbon reserves, metallogenic deposits are associated with the Cretaceous ophiolites found in Cyprus, Syria, and through Iraq to Oman, with economic copper and massive amounts of sulphides. The major mineral resources of this interval, however, are the huge phosphate deposits encircling all of the Levantine and the north-western Arabian plate and found across Egypt westward to Morocco–Mauritania. They are mostly in the Late Cretaceous marls, chalks and limestones. Their significance is twofold as some contain low-level uranium mineralization.

The Palaeocene transgression covered almost the entire Arabian platform, consisting principally of neritic carbonates (Umm ar Radhuma formation), but with open marine chalky marly facies in the Levant region. This continued in the Levant through the Eocene, but in Arabia restricted depositional conditions resulted in the Rus gypsum–anhydrite formation from southern Iraq to northern Somalia. Vertical uplift over the entire region occurred in the Late Eocene, with tensional tectonics over the Red Sea–Aden area. Magmatism was intermittent through the Oligocene and

Early Miocene with fracturing preferentially following old lines of weakness, the Red Sea apparently developing along an ancient Proterozoic suture. The eastern part reflects shoaling and isolation as represented by Mid-Miocene evaporites interbedded with clastic red beds (Lower Fars formation). The Pliocene in the east shows clastics, while to the west, in the Mediterranean and Red Sea areas, pelagics were dominant. Arabia completely separated from Africa by the end of the Miocene, and started moving north-eastward against Iran and Turkey, resulting in the folded Taurus–Zagros orogenic ranges. Plio-Pleistocene molasse was shed from the rising mountains. Further rotations were taken up by sinistral horizontal movements along the Levantine fracture (Aqaba–Dead Sea–Jordan) with resultant upfolding of the Levant to the present.

Again, prolific hydrocarbon deposits occur in this interval, notably in the Tertiary, and very significant are the metal-rich deeps of the Red Sea with their brines and muds.

AFRICA

The African geology is compiled from Furon (1963), Afyeh and Mansour (1977) and Mengoli and Spinicci (1984).

The Nubian part of the Arabo-Nubian Shield consists of basement crystalline and metamorphic rocks outcropping in southern Sinai, the Red Sea mountain chains in Egypt and through Sudan south to Somalia, with patches at Asswan and Ouweinat Mountain in south-west Egypt. There were tectonic and plutonic episodes, and faulting to the top of the sequence associated with vast outpourings of lava (Dukhan Mountain north of the eastern desert). This was followed by extensive deposits (Hammamat formation), with a return to magmatic intrusions forming many of the Red Sea mountains.

Palaeozoic sandstones are limited in north-west Sudan, south-east Libya and south-west Egypt and covered by younger sandstone deposits, although in Egypt Precarboniferous extensive carbonates thicken to the north, but are distorted with folding from the north-east to the south-west. Other micaceous sandstones (Nawa formation) are present in mid-western Sudan directly overlying Precambrian and underlying Cretaceous Nubian sandstone. The Suez subsiding geosyncline received clastic deposits in the Carboniferous, as reflected by two thick sandstones exposed at the Gulf of Suez, with a dolomitic limestone between associated with manganese nodules (Umm Bajma Mountain).

During the Mesozoic, most of Sudan was a positive area, and thus does not have many deposits, except for erosional products, while in Egypt a regression followed the Carboniferous and then a transgression in the Triassic, covering most of its northern territory. The important Nubian

110

sandstone facies characterize the Early Cretaceous in both countries, underlain by the Early Jurassic dolomitized carbonates interbedded with some clastics. Marine transgression climaxed in the Middle and Upper Cretaceous, reversed to shallow marine carbonates with some phosphates and then the deeper carbonate chalky facies of the uppermost Cretaceous. The Cretaceous witnessed some tectonic activity, explaining the discontinuous features and resulting in uplifts oriented north-east to south-west. These still show in the present as high hills in Sinai, Shabrawit and Abu Rawash.

The Cenozoic in Sudan began with freshwater deposits, while in Egypt the Palaeocene and Eocene carried on from the Late Cretaceous with transgressive but shallow marine facies in mid-Egypt, marls around the Suez Basin, and similar but thinner and more variable deposits in the north. The end of the Eocene witnessed sea regression and exposure of the Red Sea–southern Sinai mountains, as represented by the Minia and the Mukattam formations. The Oligocene was a tectonically very active period with volcanicity along large fractures in Sudan and basaltic outpourings around the Gulf in Egypt and forming the Ethiopian Plateau; this continued through the Quaternary. The Suez Gulf linked itself to the Mediterranean at the end of the Oligocene and Early Miocene. In the north, continental and fluvial thick deposits formed, while in Sudan there were thick extensive laterites. The end of the Tertiary in Sudan uplifted the Red Sea mountains and the Nubian Shield, resulting in an extensive depression in the south, middle and north-west of the country which received lots of fluvial and erosional products, accumulating thick deposits (Umm Ruwaba Group). The Miocene of the Gulf in Egypt was continental fluvial overlain by marls, gypsum and anhydrites, along the Red Sea in Sudan. These resulted from the closing of the link with the sea, which regressed completely by the Pliocene, and then again partially relinked through the Pleistocene and Recent. The Quaternary in Sudan saw erosion with accompanying extensive clayey deposits in the middle plains, while to the west, sand flats (Kardafan) covered extensive areas, and the River Nile started to form.

Precambrian rocks outcrop in Mauritania, Morocco, Algeria and Libya, forming part of the basement rocks of the African Shield. They form huge masses with thicknesses of several thousand metres, often making high mountains or occupying elevated spots. They are variably tectonized, plutonic-metamorphic sequences with partly metamorphosed clastics and limestones. The Phanerozoic and Proterozoic are normally separated by large faults trending north–south. Different episodes of tectonic events associated with magmatic intrusions are reflected by different levels of granites, along with schists, gneisses, molasses and arkoses. Most mineral occurrences (metallics) are associated with these masses, or with their

Appendix

thrust zones. Several earth movements took place in the Precambrian, which resulted in the Atlas mountains, trending east–west, and other prominent massifs, i.e. Eglab, Ahaggar and Tibesti.

In Morocco, the geology forms three regions: Al-Rif, Al-Atlas (intermediate and Greater) and the Little Atlas. The first lies on the northern Atlantic coast and extends to around 200 km with peaks getting higher to the east. It was subjected to the Alpine Orogeny, which dictated its final form, with associated volcanism. At its north are Precambrian granites, schists and gneisses being the oldest, followed southward by the folded carbonates of Triassic, Liassic and Jurassic sequences. Further south is the Rif area, occupied by Liassic and younger schists, and rarely by carbonates. Then comes the marginal Rif area, mostly made up of marly and clayey Tertiary deposits. The second region, Al-Atlas, shows flattened and upfolded sequences. Precambrian rocks are very thick (8–10 km). The area was affected by both Caledonian and Hercynian Orogenies, the latter being more prominent, resulting in folds trending east to west or north-east to south-west, along with granitic intrusions. These are overlain by continental Permian–Triassic deposits with basalts of the Liassic, Jurassic and Cretaceous that reflect some movements associated with the Liassic–Jurassic transgressive carbonates. The Cretaceous was predominantly an exposed area, notably the Mesetta parts, which ended and merged with the Tertiary 'phosphate basin'. During the Miocene shallow marine deposits were predominant to the north, and continental in the mid-country, notably after the Eocene and into the Pliocene when a new transgression occurred. Eocene movements were responsible for the mountain chains, which were modified by the Alpine Orogeny into folded belts and huge fractures responsible for the present physiography. The Atlas region can be further subdivided into the eastern and western Mesetta, the intermediate and the Greater Atlas as one goes from the mid-country west to the Atlantic Ocean, south of the Rif. The Little or Minor Atlas region further south is separated from the Greater Atlas by a huge transcurrent fault, which also separates the Atlas from the African rock groups. African basement rocks are thus south of this fault. To the north of the Little Atlas area is an extensive depression paralleling the fault and extending from the coast eastward in the Sous plains. Precambrian rocks are exposed as granitic uplifts, with frequent volcanic floods and obvious effects of the Hercynian Orogeny.

In Mauritania, Precambrian rocks cover most of the country except the coast and the north-east. Again, there are crystalline and metamorphic rocks with extensive granites to the north-west of the country. To the east, most of the country is occupied by a sandstone and marly plateau (north-east going south to the Senegal River) of Ordovician–Silurian age. Still further north and east of this plateau there are predominantly

Devonian–Carboniferous clastics and sandstones. All of the above are highly affected by the Hercynian Orogeny. Cretaceous–Eocene formations cover coastal and southern areas of the country, while the Quaternary sands cover almost one-fifth of Mauritania. The coastal plains show sabkhas with a gypsiferous crust.

The rest of North Africa, other than the Precambrian massifs already mentioned, started in the Silurian as a widespread marine transgression depositing graptolite shales and carbonates. Toward its termination and in the Devonian the Caledonian Orogeny began, resulting in such features as the Ougarta uplift, Bou Bernous arch, Tihemboka arch, and Talemzan and Nefusa highs. During the Carboniferous, and with the Hercynian Orogeny, intense folding of the Anti-Atlas mountains occurred with the above features rising and the Saharan trough taking shape. The Permian thus started with erosion of the elevated parts, and some north–south trending structures developed in the trough. Deep water or open shelf facies followed, thicker to the north and thin on the highs over a basin from Gales to Tripolitania. In the Lower and Middle Triassic, continental conditions were widespread, with rather shallow basins witnessing two rifting phases accompanied by volcanism. These shallow epicontinental seas were covered by a large transgression at the Upper Triassic, notably in Algeria and Tunisia, while further east to Libya and Egypt conditions were trending to continental with some pelagic sediments depositing in troughs. The whole area was under very shallow waters and towards the uppermost Triassic evaporites deposited. Rifting continued from the Lower Jurassic up to the opening of the north central Atlantic in the Middle Jurassic and the separation of Africa from Europe, with Africa moving east along a left transcurrent fault. Marine reefal conditions prevailed in the Upper Jurassic, notably around Tunisia but extending to Egypt. Then, and through the Lower Cretaceous, Africa moved counterclockwise away from Europe, inducing tectonic instability and the onset of Moroccan flysch deposits along the transcurrent fault.

By the Cretaceous, the spreading of the North Atlantic ceased and Africa reversed its movement, with the closure of Neo Tethys in the Upper Cretaceous and widespread tectonic events, notably in the Betic–Rif–Tell area (Spain and Morocco) with effects from Constantine through Tunisia, Libya and Egypt. In Libya, erosional activities partially filled the Hammada Basin while there were deep sea conditions in Cyrenaica. Some uplifts collapsed in the Upper Cretaceous with the formation of grabens, erosion and thick clastic depositions in troughs. In Egypt, it was a time of general regression, while in Algeria and Tunisia reefs and evaporites formed, and in Libya the Nefusa high rejuvenated and Jabel Al-Akhdar uplifted.

The closure continued through the Palaeocene and Upper Eocene, leading to a collision between the Alboran Kabylie and the African margin, resulting in the Kabylie nappes and the structural setting of the Atlas. In the Palaeocene the Saharan platform emerged completely, with pelagic sediments from the Tunisian trough to the Gabes–Tripoli Basin, while lagoonal deposits formed in the Gafsa Basin. The Sirte Basin sank, with occasional carbonate banks interfingering with shales. The Upper Oligocene opened the north Algerian Basin between Gibraltar and Calabria, with the deposition of arenceous micaceous flysch up to the Lower Miocene. Compressional phases followed with overthrusting in the Rif-Betic, Algeria and northern Tunisia.

BIBLIOGRAPHY

Entries marked by an asterisk are in Arabic.

Aba Yazid, O. (1985). Transfer of technology, its adaptation and implanting in the Arab countries. UNESCO Regional Office for Science and Technology for the Arab States, *Bulletin*, **13**, 36–53.

*Abdel Nour, D. (1981). Industrialization in the Arab World, its present and future. *Sho'oun Arabiyah*, no. 10, December, 168–83.

*ABEDA (1982). The cooperative Arab program for development in Africa. The Arab Bank for Economic Development in Africa, a statistical report. *Sho'oun Arabiyah*, no. 12, February, 173–83.

Abed, G.T. (1983). Arab financial resources: an analysis and critique of present deployment policies. In *Arab Resources*, Ibrahim, I (ed.). London: Croom Helm, pp. 43–70.

Acharya, S.N. (1979). Incentives for resource allocation: a case study for Sudan. World Bank working paper no. 367, Washington DC, USA.

*Afyeh, M.S. and Mansour, A.O. (1977). Development of mineral resources in the Arab Nation. Cairo (now Tunis): ALECSO-ACID.

AGID (1977). New directions in mineral development policies. Report of the Bagauda International Workshop, September 1975. Bagauda, Nigeria: Association of Geoscientists for International Development.

AGID (1978). Mineral resource management in developing countries: state participation, private enterprise or both? Report of Symposium, 17 August, University of Sydney.

AGID (1985). Area selection and management planning of mineral exploration in developing countries. *Association of Geoscientists for International Development Newsletter*, no. 45, 9–12.

*Alami, A.A. (1982). The economics of sulphur in the Arab World. *Arab Mineral Resources Journal*, **2**, 87–111.

*Al-Khatib, O. (1984a). The Arab nation till the year 2000: forecasting the political conditions. *Al-Baheth*, **6**, nos 33/34, 19–54.

*Al-Khatib, O. (1984b). The Arab nation till the year 2000. *Al-Baheth*, **6**, nos 35/36, 39–58.

*Al-Manjara, Al-M. (1983). The Great Arab West in the year 2000. *Al-Mustakbal Al-Arabi*, **53**, no. 7, 4–17.

*Al-Mshat, A.M. (1983). The Arabs and Africa. Briefings on a seminar held in Amman, Jordan, 25–29 April. *Al-Mustakbal Al-Arabi*, **53**, no. 7, 129–34.

Bibliography

*Al-Olabi, M.A.M. (1987). *Program of Total Arab Developmental Needs—The Role of Science and Technology*. Damascus: ALECSO. 80 pp.

Al-Sadik, A.T. (1985). National accounting and income illusion of petroleum exports: the case of the Arab Gulf Co-operation Council Members (AGCC). In *Prospects for the World Oil Industry*, Niblock, T. and Lawless, R. (eds). London: Croom Helm, pp. 86–115.

*Al-Sammak, A.S.M. (1984). The geopolitical weight of major population patterns in the Arab nation. *Al-Mustakbal Al-Arabi*, **67**, no. 9, 92–115.

*Al-Wahda (1985). Arab society: where to? *Al-Wahda*, **1**, no. 6, 3–80.

*Al-Wattari, A.A. (1988). The role of downstream industries in Arab economic integration. *OAPEC Monthly Bulletin*, **14**, no. 2, 12–14.

*AMC (1980*a*). Mineral news of the Arab nation and the world. *Newsletter of AOMR*, no. 1, 30–1.

*AMC (1980*b*). Mineral news of the Arab nation and the world. *Newsletter of AOMR*, no. 2, 23–4.

*AMC (1980*c*). Mineral news of the Arab nation and the world. *Newsletter of AOMR*, no. 3, 16–22.

*AMC (1981). The trade status of major Arab mineral resources and their prospects. Arab Mining Company report to Fourth Arab Conference on Mineral Resources.

*AMF (1985). *The Unified Arab Economic Report*. Kuwait: Arab Monetary Fund (in co-operation with the Arab League, AFESD, OAPEC).

*AMF (1986). *The Unified Arab Economic Report*. Kuwait: Arab Monetary Fund (in co-operation with the Arab League, AFESD, OAPEC).

*AMM (1985). News section. *Arab Mining Magazine*, **5**, no. 1, 32–3.

*Ammar, H. (1982). On Arab higher education and development. *Al-Mustakbal Al-Arabi*, **40**, no. 6, 119–38.

*Anon. (1983). Lead in the Arab World. *Newsletter of AOMR*, no. 14, 28–31.

*Anon. (1985). Activities of the General Establishment of Geology and Mineral Resources (GEGMR)—Syria. *Syrian Journal of Geology*, **8**, 104–6.

*AOID (1979). About technology transfer issues in the third World. *Arab Organization for Industrial Development Magazine*, no. 37, January, 4–24.

*AOID (1985). The Sixth Industrial Conference—move to action. *Arab Organization for Industrial Development Magazine*, no. 5, January, 108–24.

*AOMR (1981). The strategy for developing mineral resources of the Arab World as an economic continuum. Arab Organization for Mineral Resources report to the Fourth Arab Conference on Mineral Resources, Amman.

*AOMR (1985). Main working paper, Fifth Arab Conference on Mineral Resources, Khartoum, Sudan.

*Arab Mining Company (1981). Current trade of major Arab mineral ores and its development. Amman: AOMR.

*Assaf, M.A.M. (1982). The problem of effective policies in the Arab countries (a theoretical comparative framework). *Sho'oun Arabiyah*, no. 12, February, 7–28.

Bibliography

*Azar, W. (1985). Phosphates in Jordan. *Arab Mining Magazine*, 5, no. 1, 5–9.

*Azzam, H. (1984). Developing manpower in the Arab Libyan Jamahiriya. *Al-Mustakbal Al-Arabi*, 67, no. 9, 116–27.

Bartocha, B. (1981). Co-operative programmes in science and technology in Arab countries. STPAS, ECWA, E/ECWA/NR/SEM.3/18.

Ben Shanho, A.L. (1981). Foreign institutes and technology transfer to the Algerian economy. STPAS, ECWA, E/ECWA/NR/SEM.3/4.

Beydoun, Z.R. (1988). *The Middle East: Regional Geology and Petroleum Resources*. Beaconsfield, UK: Scientific Press.

BGS (1988). *World Mineral Production 1983–87, Preliminary Statistics*. London: British Geological Survey.

Bhatt, V.V. (1981). Project evaluation criteria and technology policy. STPAS, ECWA, E/ECWA/NR/SEM.3/12.

Bonar, L.G. (1982). Mineral market research and strategy. In *Mineral Policy Formulation*. Canada: CRS, pp. 93–9.

*Bou Kamra, H. (1982). The role of education in developing the Arab self. *Al-Mustakbal Al-Arabi*, 40, no. 6, 105–18.

*Bsisso, F.H. (1983). Co-operative development among the Arab Gulf States. *Al-Mustakbal Al-Arabi*, 55, no. 9, 46–67.

Butterfield, D. and Kubursi, A.A. (1981). Investment planning and industrialization in the Syrian Arab Republic: a simulation exercise. *Journal of Industry and Development*, 6, 65–87.

Carman, J.S. (1977). Forecast of United Nations mineral activities. In *New Directions in Mineral Development Policies. AGID Report No. 3, part C—International Co-operation and Aid*, pp. 146–9.

Chender, M. (1986). Strategic options for mining companies. *AGID Newsletter*, no. 49, 9–15.

Cody, J., Hughes, H. and Wall, D., eds (1980). *Policies for Industrial Progress in Developing Countries*. New York: Oxford University Press.

Corm, G. (1981). Technology and investment decision-making in national financial institution. STPAS, ECWA, E/ECWA/NR/SEM.3/15.

*Corm, G. (1983). The Arabs and the First World—the trans-national companies. *Al-Mustakbal Al-Arabi*, 48, no. 2, 25–37.

Crowson, P. (1984). *Mineral Handbook 1984–1985*. London: Macmillan.

CRS (1982). Mineral policy formulation: the role of scientific and technical knowledge. Proceedings of the Tenth Center for Resource Studies Policy Discussion Seminar, Kingston, Ontario, 22–24 June.

*Dajani, B. (1982). Economic implications of the Eleventh Arab Summit Meeting. In *Studies in Economic Arab Complementarity and Development*. Beirut: Arab Unity Studies Centre, pp. 197–230.

Delhawi, M.R. and Laurent, D. (1983). Industrial minerals and building materials developments, Saudi Arabia. *Proceedings Fifth Industrial Minerals Congress*, pp. 135–46.

117

Dempsey, M., ed. (1983). *'Daily Telegraph' Atlas of the Arab World.* London: Nomad.

Desforges, C.D. (1984). The problem of precious metals and other non-ferrous metals in relation to indigenous supplies. In *Indigeneous Raw Materials for Industry.* London: Metals Society, pp. 128–33.

*DGMRS (1984). Mineral resources in Sudan. National paper presented to the Fifth Arab Conference on Mineral Resources.

*Dimachkieh, M. (1984). Problems and trends of the Arab iron and steel industry. *Al-Mustakbal Al-Arabi,* 65, no. 7, 128–40.

DMMR (1982). *Saudi Arabia Mineral Resources Annual Report 1981–1982.* Jeddah: Deputy Ministry for Mineral Resources.

DMMR (1983). *Saudi Arabia Mineral Resources Annual Report 1982–1983.* Jeddah: DMMR.

EC (1987). The European Community and the Third World. *European Community Newsletter,* no. 10.

ECWA (1977). Survey report on the situation pertaining to the development of mineral resources in the countries of the ECWA region. UN Economic and Social Council, Economic Commission for Western Asia, E/ECWA/NR/1/Rev. 1, Beirut.

ECWA (1981). Report on improvement of national codes and assessment of the situation with regard to full sovereignity of member countries over their mineral resources. E/ECWA/NR/11, Beirut.

ECWA (1983). Survey and economic analysis of the actual and potential development of industrial mineral deposits in the ECWA region. E/ECWA/NR/83/2, Beirut.

Edwards, L.L. (1983). Case studies in technology transfer in the Arab World. In *Arab Resources,* Ibrahim, I. (ed.). London: Croom Helm, pp. 185–91.

*Eid, A.M. (1977). The future of mineral industries in the Arab countries, the year 2000. *Arab Organization for Industrial Development Magazine,* no. 32, October, 115–33.

El-Fathaly, O. and Chackrian, R. (1983). Administration: the forgotten issue in Arab development. In *Arab Resources,* Ibrahim, I (ed.). London: Croom Helm, pp. 193–209.

Europa International (1985). *The Europa Yearbook 1985, a World Survey.* London: Europa Publica.

*Farid, A.M. (1982). Turkey and the Arabs. Briefings on a seminar held at Durham University, England, 14–15 January. *Al-Mustakbal Al-Arabi,* 45, no. 11, 158–69.

Financial Times (1986). *1985 Financial Times Mining International Yearbook.* Harlow: Longman.

Furon, R. (1963). *Geology of Africa.* New York: Hafner.

Galer, G.S. (1985). Scenario planning in the context of international energy development. In *Prospects for the World Oil Industry,* Niblock, T. and Lawless, R. (eds). London: Croom Helm, pp. 74–82.

*GEGMR (1984). Mineral resources in the Syrian Arab Republic. National paper presented to the Fifth Arab Conference on Mineral Resources.

*Gilliam, L.B. (1987). Technical institutes in the Arab World: filling a vital need. *Developments in MENA*, no. 59.

*Girgis, M., ed. (1984). *Industrial Progress in Small Oil-exporting Countries—The Prospect for Kuwait.* Harlow: Longman.

Haglund, D. (1983). Strategic minerals and Canada. *Center for Resource Studies Perspectives*, no. 17, 1–3.

Häfele, W. (1980). World regional energy modelling. In *World Energy Issues and Policies*, Mabro, R. (ed.). New York: Oxford University Press, pp. 187–211.

*Hashimi, B.M.D. (1984). Industrialization of the Arab Peninsula: suggestions on developing private enterprises. *Al-Mustakbal Al-Arabi*, **70**, no. 12, 100–17.

Hautala, P.C. and Hoskins, J.R. (1985). Communication: the missing link in the minerals industry. *Mining Engineering*, **37**, no. 1, 25–7.

Hufshmidt, M.M. and Carpenter, R.A. (1982). Natural systems assessment and benefit–cost analysis for economic development. Proceedings First International Symposium of the AGID/AIT, Bangkok.

Humphreys, D.S.C. (1984). Consumption of non-fuel minerals: trends and economic appraisal. In *Indigenous Raw Materials for Industry*. London: Metals Society, pp. 49–56.

*Ibrahim, S.D. (1983). Arab–European co-operation. Briefings on a seminar held in Lauvane University, Belgium, 2–4 December 1982. *Al-Mustakbal Al-Arabi*, **48**, no. 2, 147–51.

IBRD (1983). *International Bank for Reconstruction and Development Report.* Washington DC: The World Bank.

Industrial Minerals (1978). Magazine of the Metal Bulletin Group, England. Nos 131 and 133.

Industrial Minerals (1979). Nos 136, 137, 139 and 140.

Industrial Minerals (1980). No. 148.

Industrial Minerals (1983). Mining in jeopardy. No. 185, 48–9.

*Jabr, F.S. (1980). *Technology and Mining Industries in the Arab Nation.* Baghdad: Ministry of Culture and Information.

*Jalal, F. (1985). Acquiring and nationalizing effective and convenient technological capability. *AOID Magazine*, **5**, 78–107.

Kettani, M.A. (1985). Science and technology and the Muslim World. Proceedings Co-ordinating Conference on Technology in Islamic Countries, Istanbul, 21–25 October, Volume I, pp. 10–73.

Khaled, H. (1985). Fertilizers industry in the Arab World. *Arab Mining Magazine*, **5**, no. 1, 36–42.

Khawlie, M.R. (1983a). An overview of mineral resources development in the Arab World. Proceedings Fifth Industrial Minerals International Congress, Madrid, 1982, pp. 147–57.

Khawlie, M. (1983*b*). Raw materials and the mineral industry of Lebanon. Proceedings First Jordanian Geological Conference, Amman, 6–8 September 1982, pp. 373–91.

Khawlie, M. (1986*a*). The policy for small-scale mining in a developing country: Lebanon. *Resources Policy.*

*Khawlie, M. (1986*b*). Arab co-operative work: mineral resources in development. *Sho'oun Arabiyah*, **46**, 107–27.

Khawlie, M. and Attiyeh, F. (1986). War and the ceramics industry in Lebanon: a lesson to learn. *Industrial Minerals Journal*, no. 229, 59–61.

Khawlie, M. and Hinai, K. (1980). Geology and production of construction material resources of Lebanon: a preliminary study. *Engineering Geology*, **15**, 223–32.

Khawlie, M. and Khanamirian, V. (1985). The 'Black Sands', a titanium–iron ore deposit for Lebanon. *Lebanese Science Bulletin*, **1**, no. 2, 149–62.

Kubursy, A.A. (1985). Industrialization in the Arab states of the gulf: a Ruhr without water. In *Prospects for the World Oil Industry*, Niblock, T. and Lawless, R. (eds). London: Croom Helm, pp. 42–65.

Landsberg, H.H. and Tilton, J.E. (1986). The global slump in metals. *AGID Newsletter*, no. 46, 4–7.

*Lehzami, S.M. (1980). *Industrial Integration in the Arab West: An Economic Analysis*. Tunisia: Dar Al-Balagh Al-Arabiyah. 366 pp.

Lenti, R.T. (1986). Raw materials: suggesting a way out. *AGID Newsletter*, no. 48, 4–8.

Livingstone, I. (1984). Resource-based industrial development: past experience and future prospects in Malawi. *Journal of Industry and Development*, **10**, 75–135.

Maksoud, K. (1981). Industrial science and technology policies in the Egyptian iron and steel sector during the period 1950–1980. STPAS, ECWA, E/ECWA/NR/SEM.3/20.

Marzouk, M.S. (1984). Forecasting industrial output and employment until 1990. In *The Prospect for Kuwait*, Girgis, M. (ed.). Harlow: Longman, pp. 155–82.

*Masha'l, M. (1977). Encouraging and co-ordinating industrial investment among the Arab countries. *Arab Organization for Industrial Development Magazine*, no. 32, October, 5–22.

Mengoli, S. and Spinicci, G. (1984). Tectonic evolution of north Africa (from Algeria to Sinai). OAPEC Seminar on Source and Habitat of Petroleum in the Arab Countries, Kuwait, 7–11 October.

Metals Society (1984). Indigenous raw materials for industry. Proceedings Conference on Materials Forum, Metals Society, London.

Michalet, C.A. (1981). Transfer of technology by TNCs: traditional versus new forms. STPAS, ECWA, E/ECWA/NR/SEM.3/19.

*Mu'awad, J.A. (1983). The problem of political sharing in the Arab nation. *Al-Mustakbal Al-Arabi*, **55**, no. 9, 108–19.

*Muharram, M.R. (1984). *Arab Mineral Resources: Possibilities of Development in a Unified Framework*. Beirut: Arab Unity Studies Centre.

*Murad, A.F. (1982). Evaluating inter-Arab projects as an introduction to economic Arab integration. In *Studies in Development and Economic Arab Integration*. Beirut: Arab Unity Studies Centre, pp. 231–52.

*Nassar, A. (1982). *Arab Potentialities*. Beirut: Arab Unity Studies Centre.

Nyrop, R.F., ed. (1985). *Saudi Arabia—Country Study*. Area Handbook Series. Washington DC: American University.

*OAPEC (1987). Annual Report. The Secretary-General, M.S. Al-Oteiba, Kuwait.

Oldham, C.H.G. (1981). Technology policy research and the Arab States. STPAS, ECWA, E/ECWA/NR/SEM.3/13.

O'Neil, T. (1985). Minerals lag general economic recovery in 1984. *Mining Engineering Magazine*, **37**, no. 5, 379.

Paye, J.C. (1985). Encouraging a Third World recovery. *OPEC Bulletin*, **16**, no. 3, 3–7 and 16.

Radetzki, M. and Zorn, S. (1979). *Financing Mining Projects in Developing Countries*. London: Mining Journal Books.

*Rashīd, A.W.H. (1984). The influencing factors for the success of joint Arab industrial projects. *Al-Mustakbal Al-Arabi*, **65**, no. 7, 107–27.

ROSTAS (1985). Highlights of the Unesco programme and budget for 1986–1987. UNESCO Regional Office for Science and Technology for the Arab States, *Bulletin*, **13**, 23–7.

*Sa'deddine, I., Nassar, A., Abdullah, I.S. and Abdul-Fadhil, M. (1982). *Arab Future Scenes*. Beirut: Arab Unity Studies Centre.

SAMR (1982). Annual Report of the Deputy Ministry for Mineral Resources, Saudi Arabia, no. 1401–2.

SAMR (1983). Annual Report of the Deputy Ministry for Mineral Resources, Saudi Arabia, no. 1402–3.

*Sayegh, Y. (1982). Arab development and the critical triangle. *Al-Mustakbal Al-Arabi*, **41**, no. 7, 6–19.

Sayegh, Y.A. (1983). A new framework for complementarity among the Arab economies. In *Arab Resources*, Ibrahim, I. (ed.). London: Croom Helm.

*Sayegh, Y. (1984). Arab economic horizons in the eighties. *Al-Mustakbal Al-Arabi*, **65**, no. 7, 76–106.

*Sayegh, Y.A. (1985). *The Determinants of Arab Economic Development*. Beirut: Arab Institute for Research and Publishing (also in English by Croom Helm, London).

Serageldin, I., Socknat, J.A. and Birks, J.S. (1983). Human resources in the Arab World, the impact of migration. In *Arab Resources*, Ibrahim, I. (ed.). London: Croom Helm, pp. 17–35.

Shaw, P.R. (1983). *Mobilizing Human Resources in the Arab World*. London: Kegan Paul.

Steele, I. (1988). New materials threaten Third World exports. *AGID Newsletter*, no. 55, 24–5.

Stermole, F.J. (1983). Mineral investment 1983 depends on prices. *Mining Engineering Magazine*, **35**, no. 2, 128–30.

Strauss, S.D. (1983). The current downturn's impact on the mining industry. *Mining Engineering Magazine*, **35**, no. 2, 123–4.

*Taher, Th. (1981). The Arab Mining Company experience in Arab mining activities. *Arab Mining Magazine*, **1**, no. 3, 5–8.

*Ta'mullah, K. (1982). Demographic evolution in the Arab nation. *Sho'oun Arabiyah*, no. 11, January, pp. 106–25.

UN Bulletin (1985 and 1989). Monthly Bulletin of Statistics. Dept of International Economic and Social Affairs, Statistical Office, New York, Vol. XLII, 1988, Vol. XLIII, 1989.

UNIDO (1979). *Industry 2000—New Perspectives*. Vienna: United Nations Industrial Development Organization.

UNIDO (1981a). Industrial carrying capacity and the Lima target. *Journal of Industry and Development*, **6**, 19–35.

UNIDO (1981b). Modelling the attainment of the Lima target: the LIDO model. *Journal of Industry and Development*, **6**, 20–36.

UNIDO (1981c). *Industrial Processing of Natural Resources*. Vienna: United Nations Industrial Development Organization.

UNIDO (1982). *Handbook of Industrial Statistics*. Vienna: United Nations Industrial Development Organization.

UNIDO (1983). *Industry in a Changing World*. Special issue of the Industrial Development Survey for the Fourth General Conference of UNIDO. Vienna: United Nations Industrial Development Organization.

UNITAD (1981). The UNITAD project: a world model to explore institutional changes over the long run. UNIDO–UNCTAD team. *Journal of Industry and Development*, **6**, 37–65.

United Nations (1984a). Commodity trade statistics 1982 (Morocco). Statistical Papers, Series D, Vol. XXXII, nos 1–21, pp. 155–201.

United Nations (1984b). Commodity trade statistics 1982 (Saudi Arabia). Statistical Papers, Series D, Vol. XXXII, nos 1–15, pp. 146–243.

United Nations (1984c). Commodity trade statistics 1982 (Jordan and UAE). Statistical Papers, Series D, Vol. XXXII, nos 1–22, pp. 235–88 Jordan, pp. 140–98 UAE.

United Nations (1984d). Commodity trade statistics 1982 (Egypt). Statistical Papers, Series D, Vol. XXXII, nos. 1–20, pp. 207–67.

Urquidi, V.L. (1983). The South has no other way than self-reliance. *AGID News*, no. 37, pp. 1–6.

US Bureau of Mines (1979). *Status of the Mineral Industries—Mining, Minerals, Metals, Mineral Reclamation*. Washington DC: Bureau of Mines, US Dept of the Interior.

Bibliography

US Bureau of Mines (1983). *Minerals Yearbook*, Vol. III Area Reports. International. Washington DC: USDI.

Uytenbogaardt, W. (1977). The need for justified management of world mineral resources: a global view. In *New Directions in Mineral Development Policies*. AGID, pp. 32–42.

Van Rensburg, W.C.J. (1986). *Strategic Minerals*, Vol. 1. New York: Prentice-Hall.

Wilcock, H., ed. (1987). *Financial Times Mining International Yearbook 1988*. Harlow: Longman.

*World Bank (1979). *World Development Report*. Paris: Arab Press.

World Bank (1984). World Development Report. New York: Oxford University Press.

*Yamut, A.H. (1984). *The Significance of Industrialization in Arab Development*. Economic Studies Series. Beirut: Institute of Arab Development.

*Zahlan, A. (1981). *Technological Implications of Arab Unity*. Beirut: Arab Unity Studies Center.

*Zalzala, A.H. (1982). Economic Arab complementarity facing the challenges. In *Studies in Economic Arab Complementarity and Development*. Beirut: Arab Unity Studies Center, pp. 135–58.

*Zurayk, C. (1982). Demands of the Arab future. *Shou'un Arabiya*, no. 11, January, 8–18.

INDEX

Page numbers in *italics* refer to figures; those in **bold** refer to tables. Some textual matter may occur on the same page.

125